AF541581

Hill Agriculture and Sustainability

About the Author

Dr N.P. Thakur is Doctorate in Soil Science and Agricultural Chemistry, presently working as Chief Scientist at AICRP-IFS has made significant contribution in the field of research, Extension and teaching of soil science/FSR and worked on different aspects of soil fertility, soil salinity , sodicity and their fertility management and reclamation with special reference to forestry. Also worked on crop diversification, IPNM,,SSNM ,developed a suitable IFS model for livelihood improvement of small and marginal farmers under irrigated areas of Jammu region. Further edited 2 books, 5 book chapters,50 research papers, 10 popular articles, 01 bulletin, 02 concept papers and presented more than 40 papers in various seminars and symposia.

Dr Manpreet Kour is Doctorate in Agronomy, presently working as Assistant Professor (Senior scale) Division of Agronomy, SKUAST Jammu and have teaching, research and extension experience of about 17 years. She has teaching experience of more than 150 credit hours and handled 07 research projects, Her publications include 33 research papers, 47 abstracts and 68 extension publications along with 08 manuals, 08 popular articles, 34 chapters in book/ compendium, 02 research bulletins and presented more than 20 papers in various seminars and symposia. Apart from these 10 recommendations and technologies have been developed. Her contribution has been recognized by 14 awards/ appreciations.

Dr. Rayees Ahmad Shah is Doctorate in Agronomy. He served in Department of Sheep Husbandry (2005 to 2007) in Department of Agriculture & Farmers Welfare (2007 to 2008) in the Sher-e-Kashmir University of Agricultural Sciences and Technology of Kashmir (2008 to 2013) and in the University of Kashmir, Hazratbal Srinagar since 2013. He has to his credit a number of research papers published in National and International scientific journals. Dr. R. A. Shah authored six books and also edited five books Dr. Shah has been honored with Young Scientist Award, Best Paper Presentation Award, Best Ph.D. Thesis Award and Outstanding Achievement Award. Dr. Shah is member of many professional societies having 18 years of experience in the field of agricultural science and extension..

Dr. Parshotam Kumar is Doctorate in Agronomy presently working as Technical Assistant at AICRP-IFS has made significant contribution in the field of research and having experience of about 17 years. His publications include 20 research papers 30 abstracts and 25 extension publications along 01 popular articles, 04 chapters in book/ compendium, and presented more than 8papers in various seminars and symposia. Apart from these he has developed 08 technologies.

Hill Agriculture and Sustainability

N.P. Thakur
Manpreet Kour
Rayees Ahmad Shah
Parshotam Kumar

FU 75, Pitam Pura
New Delhi 110034
publishmywork23@gmail.com
Mob. 91-8447075807

ISBN: 978-81-96458-80-5

Preface

From the past most of the farming development efforts were made in the hills were based on the poor understanding of the hills, their resources, environment and the socio activities of the hill communities. Hill agriculture is dominant by inherent constraints, water scarcity, fragmented land holdings, poor input availability, loss of soil fertility and erosion.

The present publication entitled 'Hill Agriculture and Sustainable' comprises of seven chapters. The first chapter is about description of several niche areas of hills that have comparative advantage for better exploitation of resources and for better trade. Second chapter is about dominant features of hill farming, third bares the problems in hill agriculture where as fourth ,fifth , sixth chapters indicates possible pathways through alternate farming, soil conservation, water conservation measures, whereas seventh chapter dealt with development of modules of integrated farming system for hills .

Authors extend their sincere thanks and indebtness to the distinguished authors of various standard texts books and other publications or sources from which most of the material or literature have been drawn. In several cases it has not been possible to obtain permission for reproduction, for which authors and publishers offer their apologies. If any errors may kindly be pardoned and brought to the notice for further improvement.

Deeply felt gratitude is due to the Hon,ble Vice chancellor SKUAST-Jammu for inspiration and guidance. Sincere thanks are also due to colleagues and friends for their constant help and encouragement.

We hope this present publication will be useful to students and academicians working in agriculture and allied sciences.

Authors

Contents

1

An Introduction to Hill Agriculture

Mountains play a vital role in sustaining about 10 per cent of the world population directly as they are the major source of water supply and majority of rivers originate from these ecosystems thus sustain the life of people residing in the plains. In India hill and mountainous areas are vastly distributed all over the country with the major mountain ranges in the Himalayas and the Western Ghats. The Himalayas are extending in 2500 km length and 250 to 400 km breadth (Khanday *et al.* 2004). They are further classified in three major categories i.e **Western Himalayas** (including Jammu and Kashmir and Himachal Pradesh), **Central Himalayas** (including eight hill districts of Uttarakhand), **North Eastern Himalayas** (including Sikkim, Manipur, Meghalaya, Nagaland, Tripura, Union Territory of Arunachal Pradesh, Mizoram, hill areas of Assam and Darjeeling district of West Bengal). Whereas **Western Ghats** also known as **Sahyadri** including border of Gujarat, Maharashtra, Goa, Karnataka, Kerala and Tamil Nadu .These hilly regions are inhabited by 51 million people, covering 18 per cent of the geographical area and 6 per cent of Indian population (Wani, 2011)but hilly regions are facing crisis and are constantly struggling with their underdevelopment and under pressure from natural and human-induced stresses. In the last few years in India equitable growth getting more importance and thus there has been a significant shift in the focus of economic policy . In view of the potential of hills for economic growth, development of rural sector within the hill region, and their role in providing life sustaining water and environmental services, hill agriculture has started receiving attention of government and other voluntary agencies. The development of hill and mountain areas and protection of their ecology have become matters of national concern in recent years emanating from the fact that hills and mountains are the recognized climate markers .The increasing population pressure is further likely to strain the available resources of plains limiting their capability in meeting the future requirements.This means there should be inclusive growth and benefits of growth is to spread the to all sections of the population and geographical regions of the country. This change in approach becomes important for the hilly regions of the country, as the rest of the economy is almost doing well.

Hill or mountain agriculture is defined as that type of livelihood system in which farmers are dependent on all land based activities to make a living, such as, cropping, horticulture, livestock, rangelands and pastures, forests etc. Due to the poor scope for industrialisation in mountains, agriculture remains an important sector for rural livelihood and economic growth inspite of its declining share in the economy. Hilly states with potential of surface water and agro-climatic diversities have a lot of potential to hasten the agricultural growth through diversification (from low to high value crops), organic farming (as most of the hills are already organic by default), agro-forestry, horticulture farming, floriculture, integrated farning system, grazing and pasture management, quality seed production also the demand for attribute based products that can be produced only in hill ecosystem is rising rapidly. Livestock sector is also an important component of various farming systems in hills as they have symbiotic relationship with crops and also provides major products like meat, milk, wool even world famous pashminas. Further, they are main source of draught power for undulating agricultural lands of hills where the possibility of mechanisation is least. Soils in hills are subject to erosion and many other natural processes which reduces their soil fertility and further productivity in this case the use of animal manure in the form of FYM and vermicompost is the vital input for restoring and maintaining the fertility of hill soils.

Conclusion

Hill agriculture has several niche areas having comparative advantage for better exploitation of resources and for better trade. These alternate type of farming offer great scope for enhancing the farm income, women empowerment and creating employment opportunities. But, this potential needs to be harnessed with great care without disturbing the natural system and also in agreement with socio-economic and cultural factors. It is suggested that interventions in the hills in the form of alternate agriculture, soil and water conservation measures and integrated farming system or zero budget natural farming the hill agriculture could produce a sustainable livelihood.

2

Characteristics of Hill Agriculture

Hilly areas of India, especially the Himalayas and the Nilgiris which are rich in their forests areas and watersheds have played a major role in maintaining climatic and ecological balance in the country for a long time. These hills and mountain areas have specific characteristics which differs them from plains in diversity of habitats for flora and fauna, ethnic diversity, topography, elevation, physiographic feature land use systems and socio-economic conditions.

Agro-biodiversity

The hills are characterized by rich agrobiodiversity. These rich agro-diversity elements have evolved, over the years due to association with the communities distributed along the mountain and hill slopes. The interactions between the tribal people and the natural system haven helped in maintaining the richness of its genetic materials in both production systems and wild lands species. Himalayas are considered as primary and secondary center of origin of rice, maize, millets, chillies (Capsicum) and cucumbers representing a rich of germplasm diversity. That large number of quality rice of glutinous, aromatic and medicinal values are produced under the influence of micro climatic conditions and their different ecosystems in the state (Changkija 2012). From NCOF report 2005, it has been observed that the Himalayan region with diverse agro-climatic condition from subtropical plain to semi-temperate to alpine and sub-alpine conditions, inhabit 39 million people and large percentage is of hill farming communities (mountains included) (Table 1).

Table 1: The Himalayan Region; Demographic and Agriculture Indicators

State	Area (sq. kms)	Population (no)	Rural (numbers)	Population density (per sq.kms)	Area under forests ('000 ha)	Net cropped area ('000 ha)	Net cropped area as % of total area	Cropping intensity (%)	Av size of holdings per family
North Western Hill									
Himachal Pradesh	55673	6077900	5482319	109	1094	551	12.16	174	1.16
Jammu &Kashmir	222236	10143700	7627062	46	2747	733	16.27	147	0.76
Uttarnchal	53483	8489349	6310275	159	3342	788	14.91	164	1.01
Total	331392	24710949	19419656	75	7183	2072	14.47	160	0.97
North Eastern Hill									
Arunachal Pradesh	83743	1097968	870087	13	5154	166	3.02	159	3.31
Assam	78438	26655528	23216288	339	1930	2701	34.41	152	1.17
Manipur	22327	2388634	1818224	107	602	140	6.33	142	1.22
Meghalaya	22429	2318822	1864711	103	938	240	10.71	111	1.33
Mizoram	21081	888573	447567	42	1599	91	4.31	100	1.29
Nagaland	16579	1988636	1635815	119	875	261	16.73	113	4.82
Sikkim	7096	540851	480981	76	257	95	13.38	127	1.65
Tripura	10486	3199203	2653453	305	606	277	26.41	152	0.6
Total	**262179**	**39078215**	**32987126**	**149**	**11961**	**3971**	**17.09**	**145**	**1.92**
India	3,287,240	1028,830,774	742706609	312.98	69024	141231	46.15	134	1.41

Source:(NCOF Vol.II. Report , 2005)

Due to wide diversity Himalayan regions are suited to different crops e.g. apples in Himachal , safforon in Kashmir,Pashmina in ladakh or mithun in Arunachal Pradesh. All this offers hope to promote investment in such areas and boost farm economy in sustainable basis (Partap,1995; Partap,1998)

Soil resources

Soils of the hills and mountains are defined to be those found in areas with altitude above 300 m above sea level and slope of 18% and above (Carating *et al.,* 2014). In india Mountain soils are found mainly in Jammu and Kashmir, U.P., West Bengal in the Himalayas submontane tracts. These mountains and hills in India show a variety of soils

Light to heavy soils having moderate to poor fertility are found in the soils of outer plains of **Jammu and Kashmir**, whereas alluvial soils with sufficient organic matter and nitrogen content are found in Kashmir region. Brown soils deep and rich in humus good for orchard crops are found in outer hills extending upto lower fringes of middle mountain region covering deciduous forest belt. Spodic soils are most extensive in the middle mountain region and cover vast areas in Jammu and temperate zone of Kashmir region. The soils of the **Himachal** can broadly be divided into nine groups on the basis of their development and physico- chemical properties. These are: (i) alluvial soils, (ii) brown hill soil, (iii) brown earth, (iv) brown forests soils, (v) grey wooded or podzolic soils, (vi) grey brown podzolic soils, (vii) planosolic soils, (viii) humus and iron podzols (ix) alpine humus mountain speletal soils. In **Uttarakhand** dark grey, grayish brown to brown colour soils are found which are generally non-calcareous and neutral to slightly acidic in reaction. In high rainfall areas moderate to highly acidic soils are found due to leach down of the bases from the soil minerals.Under the temperate climatic conditions brown forest soils, sub-mountain soils, mountain meadow soils, skeletal soils and red loam soils are the major soil groups found in Uttarakhand hills. Podzol soils or brown forest soils good for growing maize, barley, wheat and fruits are found in coniferous forest belts of Jammu and Kashmir, Himachal Pradesh, Uttarakhand, Sikkim. Whereas, Alpine Meadow soils are found in the Alpine Zone of the Himalayas.

Climate

Himalayas are divided in to four groups on the basis of climatic conditions viz. sub-tropical, sub-temperate, wet temperate and dry temperate. The average rainfall in this region varies from > 250 mm in dry temperate to 2800 mm in sub-humid zones. In the southern foothills the average summer and winter temperature is about 30 °C and 18 °C, respectively. In the middle Himalayan valleys the average summer temperature is around 25 °C while the winters are very cold. On the higher region of the middle Himalayas the summer temperature is recorded at around 15 to 18 °C while the winters are below freezing point. The regions above 4880 m are permanently covered with snow and their temperature is below freezing point.

Water sources

Water in the hills and mountainous areas become available through precipitation in the form of rains, sleet, hails and snow. These runoffs from the hills and melting of snow from high hills turn into river streams. Apart from snow meltwater, these hills are also benefitted from rainstorms in spring, and from the abundant monsoon rainfalls during the summer. Thus ground water recharge and melting of snow are natural reservoirs which also give rise to perennial sources of rivers and springs .

Rainfed farming

Also the dominant features of hill farming is that they have rainfall-dependent farming. The summer crop season receives approximately about 75% of the total annual rainfall, of which much goes to waste through runoff (Anonymous 2008). Rainfed cereals, fruits, flowers, and medicinal plants are grown.

Shifting Cultivation

It is also a technique used in hills which refers to a rotational farming in which land is cleared for cultivation (normally by fire) and then left to regenerate after a few years. The shifting cultivation is commonly referred as Jhum cultivation in North east India as koman or Bringa in Orissa, as Kumari in Western Ghats and as Watra in southeast Rajasthan.It is also named as Sweden or slash and burn cultivation (Changkija, 2017). It is a form of low- input agriculture and fallow management and it is found common in South East Asia mostly in paddy, amaranth, maize, casa va-based system etc. previously the Jhum cycle was long and ranged from 20 to 30 years, the process worked well. But now a days , with increase in human population and increasing pressure on land, Jhum cycle reduced progressively from long years to 5-6 years causing problem of land degradation and threat to ecology of the region at large

Agriculture

Agriculture is the main source of livelihood for the hill farming communities and they practice both settled as well as age-old traditional practice the shifting cultivation (Jhum Cultivation) (Anonymous,2010). Agriculture is the main source of the economy of the hill people, contributing to 45% to the total regional income. The main crops are wheat, maize and rice, pulses, oilseeds, millets, vegetables and fruit crops. In the higher altitude , farmers grow buckwheat, saffron, black cumin, millets and grain amaranth. The major cropping systems followed are maize-wheat, rice-wheat, and the intercropping of pulses and oilseeds in maize and wheat. Further, in higher hills monoculture farming is prevalent in summer, large varieties of fruits, vegetables, spices, medicinal and aromatic plants are also grown the kharif cropping season starts with April-May and these crops are harvested in October-November. In Himachal Pradesh, potatoes, turmeric, cabbages, turnips, Chinese cabbages, lettuces or pea are grown alone, or intercropped with potato. The main fruits are apple, apricot, walnut and citrus. Potential horticultural crops for this region are plum, peas, peach, citrus and some non-perishable crops like pecan nuts, kiwi, pomegranate and olive (Fatima and Hussain, 2013). It was observed that hill agriculture is gradually diversifying in favour of fruits and vegetables. Hill states have potential for production of vegetables in off-season that has higher demand in neighbouring plains when there is scarcity of supply (Wani, 2011). Off-season vegetables, potato, ginger, turmeric, garlic and onion under irrigated conditions, and maize, rice, rapeseed, mustard, soybean, linseed, black gram and horsegram under rainfed conditions are the major crops. In a home garden, multiple crops are grown in a multi-tier canopy. The homestead has fodder trees such as *Celtis australis* (*khirak*), *Bahaunia variegata* (*kachnar*), and *Grewia optiva* (*beul*) in the upper story. The middle storey has bushes like *Adatoda vasica, Vitexnegundo*, lemon and *galgal* (*Citrus aurantifolia*) as Tea and hops are important commercial crops in the high lands of the western Himalaya (Fatima and Hussain, 2013).

Table 2: Production Trends in food grains, vegetables and fruits in the Himalayan states

States	Foodgrains		Fruits		Vegetables	
	1990-91	2003-04	1990-91	2003-04	1990-91	2003-04
N. Western Hill region						
Himachal Pradesh	80.27	75.81	14.42	20.29	3.55	3.15
Jammu &Kashmir	83.37	79.00	11.17	12.75	3.88	4.56
Uttaranchal	79.82	77.39	11.93	14.84	4.52	7.05
Regional Average	81.07	77.40	12.49	15.87	4.01	5.06
N. Eastern Hill region						
Arunachal Pradesh	76.19	68.71	8.18	15.82	6.92	7.91
Assam	71.30	65.61	1.90	2.73	5.83	5.84
Manipur	81.00	55.74	9.40	8.10	5.90	34.75
Meghalaya	55.46	49.70	10.08	9.06	10.79	13.42
Mizoram	78.35	75.27	11.77	17.27	7.59	6.18
Nagaland	83.14	66.97	2.48	7.96	3.90	8.38
Sikkim	64.54	60.32	5.07	9.76	5.00	11.27
Tripura	65.54	62.24	10.11	5.44	6.82	7.31
Regional Average	71.08	64.41	3.76	4.78	6.12	8.14
India	68.82	65.94	1.55	2.14	3.01	3.29

Source: (NCOF Vol.II. Report , 2005)

Livestock rearing

Sheep, goats, cows, mules, donkeys and crosses of cattle and yaks and backyard chicken are kept by farmers in the higher hills as a source of manure, milk meat and wool. During early April the nomadic tribes people return to their villages from lower areas to manure fields during the early growing season. After this, they gather the village livestock for summer grazing in high mountain pastures, and when summer approaches the stock migrates to still higher altitudes. In September or October sheep and goats are brought back to the lower ranges, following traditional routes.

Conclusion

The dominant features of hill farming are agro-bidiversity, shifting cultivation, rainfed farming with major rainfed cropping systems i.e maize-wheat, rice-wheat, and intercropped pulses and oilseeds in maize and wheat, while rice-wheat and vegetable based crop sequences are dominant under irrigated conditions. Livestock rearing is also an integral part of hill farming mostly done by nomadic tribes

3

Hill Agriculture and Constraints

Besides diversities and advantages in hills, the communities sustain largely on subsistence farming due to following constraints

A.Natural constraints in hills

There are large number of Inherent limitations in hill agriculture like, remote areas, vulnerability, marginality , eroded soil and short growing seasons. These Inherent constraints can be judged by the use of **unsustainability indicators of hill agriculture** which further showed increased in soil erosion rates from sloping farmlands , leaving of agricultural land due to decreasing soil fertility, appearance of stones and rocks on the cultivated lands, reduction in forest and grazing areas and more emphasis on monocropping that widened the cycle of inadequate food production and lead to food insecurity and increasing unemployment and frustration (Table 3) (Shrestha, 1992). These natural constraints are:

- **Shifting cultivation** :This current practice of shifting cultivation is found mainly in eastern and north eastern regions of India and is considered as wasteful and irrational form of land use. The bad effects of shifting cultivation includes deter orating the environment and ecology of these regions. The previously 15–20 year cycle of shifting cultivation on a particular land has lessened to 2–3 years now and has resulted in large-scale deforestation, loss in soil fertility, more invasion of weeds and other species. The indigenous biodiversity has been affected to a large extent.

- **Soil Losses:** Due to rapid and unplanned development without recognition of these unsustainability issues is destructing the natural resources of the hill areas in the form of soil loss it is mostly due to wrong way of farming practices adapted in the hilly areas without taking in to consideration the land capability classifications. Due to erosion many types of problems are foun in hills like formation of rills, gullies, ravines, landslides, torrents, siltation of reservoirs and stream beds. These gullies are scatterdly found in the hill areas. In some cases, the impact was so pronounced that the sub-surface bed-rock was exposed (Islam,1984).

- **Nutrient Losses**: nutrient losses are a major threat to farmland productivity in hills. It is mainly due to occurance of soil erosion in undulating topography of hills and also due to high rate of precipitation. The process is accelerated by open cultivation system on steep to very steep land (Farid and Hussain,1988).

Table 3: Understanding unsustainability of hill agriculture through indicators

Indicators Reflecting Problems Relating to Resource Base / Production Flow and Resource Management	Range of Changes
Soil Erosion Rates on Sloping Lands	+20 to 30 %
Abandonment of Agricultural Land due to decline in soil fertility	+3 to 11%
Appearance of Stones / Rocks on Cultivated Land	+130 to 100 %
Size of Livestock Holding per Family (LSU)	-20 to 55%
Area of Farmland per Household	-30 to 10%
8. Forest A 8 Forest Area	-15 to 85%
Pasture/ Grazing Area	-25 to 90%
Good Vegetative Cover on Common Property Land	-25 to 30 %
Fragmentation of Household Farmland (in number of parcels)	+20 to 30%
Size of Land Parcels of Families	-20 to 30 %
Distance between Farmland Parcel and Home	+25 to 60%
Food grain Production and Self- Sufficiency	-30 to 60%
Permanent Out migration of Families	None to 5%
Seasonal Migration	High to High
Conversion of Irrigated Land into dry land farming due to water scarcity	+7 to 15 %
Average Crop Yields on Sloping Lands a. Maize and Wheat b. Millets	 -9 to 15% -10 to 72%
New Land Under Cultivation	+5 to 15%
Human Population	+60 to 65%
Application of Compost (organic manure)	-25 to 35%
Labour Demand for Falling Productivity	+35 to 40%
Forestry Farming Linkages	Weak to Weak
Food grain Purchases from Shops	+30 to 50 %
External Inputs' needs for Crop Production	High to Medium
Fuel wood Fodder Scarcity in terms of time spent in collection	+45 to 200%
Fodder Supply from a. Common Land b. Private Land	 -60 to 85% +130 to 150%
Emphasis on Monocropping	High to High
Steep Slope Cultivation (above 30 %)	+10 to 15%
Weed and Crop Herbaceous Products' used as Fuel wood	+200 to 230 %
Conversion of Marginal Land into Cultivation	+15 to 40%
Fallow Periods	From 6 to 3 months

Note: A positive sign (+) means increase and negative sign (-) means decline/ decrease Time Frame of changes : 1954-1991 = 37 Years

Source ; Shrestha,1992

- **Water scarcity** It is the lack of fresh water resources to meet water demand. About 90% area of the hilly areas are rainfed and further soils of hills are undulating having steep slopes, shallow to medium in depth, having coarse to medium –textured with relatively low water retentive capacity (Miah,1993) and more runoff losses . Spatial and temporal variability in rainfall amount, frequency, intensity, distribution, and also in the onset and recede timings per se is very high. Further, The summer crop season receives about 75% of the total annual rainfall, of which much goes to waste thus further increasing the difficulties of hill communities(Anonymous, 2010)
- **Poor soil fertility** it is found in the hills as soils of these regions are shallow, gravelly and poor in fertility due to mixing of these soil with fragments of parent rocks occurring within a few centimeters, except in case of valleys or depressions where they go up to about two metres. Moreover, the large part of relatively fertile soil of hills is under the forests and hence cannot be used for agricultural purposes so agriculture takes place mostly in the valleys or scattered pieces of land on the hills that have the requisite fertility
- **Unpredictability in nature** unseasonal or heavy rainfall, hailstorms, floods, epidemic diseases, insects and an erratic monsoon, untimely snowfall sometimes completely destroy the vegetables and significant part of the output from fruit orchards and vegetable farms regularly, before they are ready for harvesting.

Socio-economic issues

Some socio –economic issues related to hills are:

- **Small and fragmented land holding:** Agriculture is the main source of livelihood for the communities residing in the Himalayan regions. For large number of small and marginal farmers of the Himalayan region, the shrinkage of cropland holdings is a major key concern for managing food and livelihoods of these agrarian communities (Partap 1998). The great majority of the land holding in the Himalayan states are less than 0.5 ha or small landholders with farms of 0.5 to 1.0 ha. Wheras, the average land holding in Jammu & Kashmir is about 0.76 ha (Table 1) These small land holdings of less than 1 ha which is further divided into a range of flat and sloping land types make difficult for farming communities to sustain their livelihoods. It has been estimated that 37% of the cropland is sloping land of various degrees in the Himalayas and these farmers are even cropping on sloping lands which are beyond 25 and 30 degrees

(Partap, 1999) and not suitable for agricultural crops. Also, down in the valleys, competition among new human settlements, urbanization, industrialisation and government infrastructure development activities, are converting the valley crop land into non farm use.

- **Low population density:** Hills are characterized with villages and scattered agricultural areas situated at great distances from roads and markets thus prevents the development of market-based institutions. Thus the scattered and small size settlements in hill areas is producing a challenge for development programmes and market formation (Nakagawa,1998;Sugaya,1998). Lack of roads, poor infrastructure and meager means of transportation reducing the population density in hills and force the hill people to immigrate to big cities.
- **Immigration:** Hills have poor employment generation capacity due to this huge numbers of young population is migrating to the cities in search of livelihood and thus there is dearth of labour during the peak agricultural demanding seasons (Gim,1998), Further, the customs of helping each other and sharing respective family labour in the labour intensive agricultural operations like transplanting ,hoeing and harvesting is also found to be in declining trend in the rural areas. Hence, in hills the farming operations are found to be completely dependent upon the women and children, which constituted 75-80% of family labour.
- **Subsistence farming:** Due to lack of capital, infrastructure, technology and market in hills, hill agriculture is dominated by subsistence level of farming i.e. self-sufficiency farming system in which the farmers focus on growing enough food to feed themselves and their entire families. Even though, there is extension of credit institutions, the interest rates on loans are found to be more in rural sector than in urban sector (viz. housing loan, car loans etc.).
- **Adoptability of the farmers:** Local strains of many agricultural commodities having better quality and taste e.g local maize, local rajmash, local vegetables etc. which sells at premium in the market, many farmers are not ready to adopt improved hybrid varieties of crops thus local strains need to be improved.
- **New generation of farmers**: Most of the educated unemployed youth across the Himalayan states belong to the farming background, no doubt, they have acquired traditional knowledge of farming from their respective families, but they no longer find it remunerative to get engaged in it and making it a source of livelihood (Sugaya,1998). So they are switching to better enterprises and profession.

Technological issues

- **Low availability of inputs and new techniques:** There is poor availability of inputs viz. quality seeds , propogated materials of improved varieties of food grains, vegetables and fruit crops, fertilizers and manures, also they are they are not available in time. From the point of view of the farmers, the major technological constraint is difficulty in availability of new varieties which have wider adaptability and resistant to drought situations, as about 90% area of the district is rainfed. The lack of knowledge i.e scientific know how of production of these crops or agronomic operations to be done in the field are also not provided to the farmers(Anonymous, 2010).

- **Agricultural Mechanization:** Due to the lack of physical facilities (viz. road networks and electricity) , undulating topography, narrow terraces and fragmented land holdings in hilly areas, hill agriculture is mainly depend upon the human and animal power . The traditional wooden tools and implements have continued to remain in use in the hills and mountains (Pariyar, 2001) and this problem is further provoked during peak periods like transplanting,weeding and harvesting when labour services get scarced .

Institutional and policy constraints

With few exceptions, constraints to improving horticulture crops in the Himalayas include inadequacy of quality planting materials, poor orchard management practices , lack of trained personnal and standard mother plants. For agricultural crops these are non-availability of the farm inputs viz. seeds, fertilizers and pesticides at right time. Less remunerative prices are fetched by the hill communities due to little access to extension services, proper marketing, post harvest processing and transport facilities (Anonymous, 2010). One of the biggest problem faced by the small farmers is the lack of an effective marketing structure, the predominant practice is to sell or to rent out the produce to middlemen much before the fruits are ready for harvest ,as a result, the farmers do not get remunerative prices from the middleman. Because of lack of regular and reliable marketing, hills farmers in most of areas, are finding it too risky to diversify into more lucrative high value crops.

Problems in animal husbandary

It includes the shortage of proper feed, fodder and concentrate to milch animals is widespread in the hills. There may be up to 70% shortage of fodder faced by the Himalayan farmers due to infestation of non palatable invasive species, such as lantana, eupatorium and congress grass in the fodder and grazing areas. In many hill regions the problem is of overgrazing due to uncontrolled herd stock and poor pasture management practices.

Due to improper market, prices for egg and broilers are not fixed, undeveloped dairy Cooperative societies, make a lot of fluctuations in the prices of milk which make the unstability in farmers income. The other problems include poor facilities of artificial insemination which decrease the breeding efficiency in animals. As no commercial hatchery exist in hilly areas these communities depend on outer sources to get day old chicks and due to high altitude (low possible oxygen pressure) cause severe stress during transportation of chicks. Lack of awareness bioscurity/insufficient veterinary aids are some of the other institutional constraints faced by the hill communities.

Conclusion

The sustainability for hill communities remain down due to natural, technological and institutional problems faced by hill communities found in the mountain agriculture besides all the diversities. The further development in hills should based farmer response, giving due consideration to the nature of marginality, fragility, diversity and other constraints of regional specific area

4

Interventions to Sustain Hill Agriculture

As already discussed in above chapter there are various physical, geographical and environmental constraints that are associated with the hill agriculture, thus the scope for` agricultural policies based on modern input-intensive agriculture is inhibited in the hilly areas resulting in the majority of the hill communities to either live on subsistence agriculture or immigrating to other parts of the country for employment. The policies that might give successful results for any other state in India situated in the plains, may not prove to be fruitful in this hilly state. As the livelihood of the weaker sections in the hill areas is absolutely dependent on natural resources, their devastation will make the process of inclusive growth unsustainable in the long run. For sustainable agriculture it is necessary that agriculture of hill can be b understood in the right way and it should include all the necessary interventions on those activities in which farmers are dependent to make a livelihood. The diagrammatic presentation of the hill specificities, along with constraints and interventions are represented below in figure-1

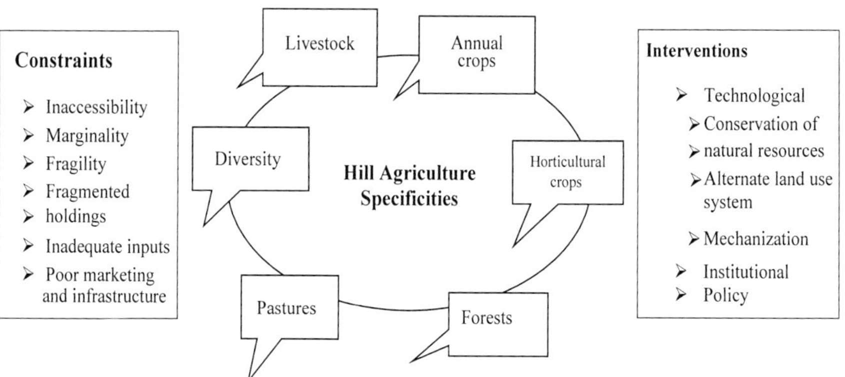

Fig. 1: Hill specificities, along with constraints and interventions

Technological interventions

Conservation of natural resources

Soil and water Conservation: soil conservation measures should always be a major consideration for sustainable hill farming as erosion causes rapid fertility depletion, damages crops by sedimentation and raising stream beds/ river channels by siltation which causes flash floods and limits discharge capacity, irrigation and navigation. The common practices in the hill tracts is to plant the crops up and down the slopes which lead to accelerated soil erosion poor rainfall infilteration with the onset of heavy rains (Pathak *et al.*, 2007). However, plantation should be made along the contours. Experiments show that contour farming alone can reduce soil erosion by as much as 50% on moderate slopes. However, for slopes steeper than 10%, other measures should be combined with contour farming to enhance its efficiency.(Gupta and gupta, 2010).

Contour cultivation

In this type of cultivation all the field operations such as ploughing, sowing or planting and intercultivation are performed on the contour. It reduces sheet and rill erosion and the resulting sediment deposition at the foot of the slope or off-site and increases water infiltration, thereby reducing the transport of nutrients and organics to surface water and increasing water storage in the soil profile.. The effectiveness of this practice varies with amount of rainfall, the soil type and topography. The Maximum effectiveness of this practice is on medium slopes and on permeable soil and it decreases with very steep slopes. This method is most effective on slopes between 2-10%. On extended slopes, where bunding is done to decrease the slope length, the bunds can act as guidelines for contour cultivation. On the mild slopes where bunding is not necessary, contour guidelines may be marked in the field (Rao *et al.*, 1981).

Other conservation measures like mulching, strip cropping, use of cover crops, adopting suitable cropping system, using conservation tillage practices, suitable crop rotation, afforestation, controlled grazing, wind breaks and adopting mechanical measure can reduce soil erosion significantly. Assessment of degraded and marginal land and their specific use is important in the hills (Ahmad,1984). A rainfed sloping farmland that is marginal for a crop because of continuous irrigation and moisture requirement could be highly productive for perennial fruit crops, may support a productive and sustainable livestock production system and herbal medicines farming (Gim 1998)

Water conservation measures The farmers of hills generally cultivate their crops under rains and the availability of surface water resources are very inadequate due to less water retention capacity of the slopy terrains of the hilly regions. Thus the rainfall is only the major source of good quality water in mountaneous areas. So, its proper management by rainwater harvesting, inter –row and inter-plot harvesting, water harvesting at surface through ponds, tanks, making earthen or masonry checkdams etc. and moisture conservation by mulching, crop covers, use of anti-transpirants, intercropping, weed control etc. need special consideration for realizing suitable crop production in these regions. Further because of the less knowledge of appropriate sowing time and other agronomic managemet practices under these rainfed cropping system the average yield is low in most cases and thus releasing of advanced package of practices can help farmers a lot to know the sowing dates.

Watershed management technology

The rain-fed agriculture contribution is 58 per cent to world's food production from 80 per cent agriculture lands (Raju *et al.* 2008). Due to increase in world population, water for food production is becoming an increasingly scarce resource, and the situation is further aggravated by climate change (Molden, 2007). In these situations watershed development program is, to be considered as an effectual tool for addressing many of these troubles and recognized as potential engine for agriculture growth and development in fragile and marginal rain-fed areas (Ahluwalia and Wani *et al.* 2006). A watershed, also called a drainage basin, hydrologic unit or catchment area, is that area in which all water that flows into it goes to a common outlet. As hills are highly susceptible to water erosion and land degradation, the nature of the soil and the inclined slopes make it difficult for ground water to collect in these areas. Thus watershed management approach becomes the most crucial factor in agricultural performance (Anonymous, 2010). It is technique of land management and stream management in a better way to improve hydrological regime. It affects the life of participated people in every sphere . Watershed leads to natural resource management and thus produces multiple benefits in terms of increasing food production, improving livelihoods, protecting environment, addressing gender and equity issues along with biodiversity concerns (Joshi *et al.* 2005 and Rockstorm *et al.* 2007). Thus productivity of food, fuel, forage, fruits and water by the management of vital resources of water, soil, vegetation and phenomenon like floods and drought are determined by nature of watershed functioning. Thus watershed approach is to ensure a holistic view of water and land resources and to prevent further degradation of these ecologically fragile areas (Gupta and Mahajan, 2010). Principals of water shed: (1) land utilization according to its capability (2) Productive top soil protection (3) reducing the hazards of siltation in storage tanks, reservoirs and lower fertile lands. (4) Covering the soil surface with the vegetation through out the year and diversion of excess rain water safely to the storage points through the vegetative water-ways. (5) Conservation of rain water in-situ (6) preventing gully erosions and stabilizing them by providing checks at specified intervals (7) increasing cropping intensity and LER through intercropping and sequence cropping (8) water harvesting for supplemental and off season irrigation (9) maintaining the sustainability of the ecosystem and maximizing agricultural productivity of the ecosystem (10) Improving socio-economic status of the farmers.

Objectives of watershed: (1) Biomass production on sustainable basis(2) Restoration of ecological balance (3) Employment generation (4) Reducing regional disparity (5) increasing the income level among the beneficiaries

The conservation of the rain water sources in hilly areas provides the key to success thus in recent years this integrated approach aims improve the standard of living of common people by increasing his earning capacity by offering all facilities required for optimum production (Singh, 2000). So watershed is recognized as the most suitable physical unit for integrated development of hill communities.

Alternate land use system

The Himalayan region is bestowed with varied landscape features that provide variety of habitats to diverse life forms ,thus hills are usually less suited for uni-dimensional land use" but more suited to multiple strategies ensuring a balanced relationship between people and land resources.

• Organic farming

Organic farming is a production system, based on renewal of ecological processes and strengthening of ecological functions of farm ecosystem to produce safe and healthy food for sustainability. It avoids or largely excludes the use of synthetically compounded fertilizers, pesticides, growth regulators, and livestock feed additives.In India especially in hilly areas where the conventional farming is still not widespread and people use local strains or depend on natural resources for soil fertility Organic Farming has strong potential. Also, most of the hill agriculture is rain-fed, the agriculture of these can be called 'Organic by default'. In India nine States have drafted organic farming policies out of which Uttarakhand, Nagaland, Sikkim and Mizoram have declared their intention to go 100 percent organic. However Uttarakhand and Sikkim has declared themselves as 'Organic States' (Meena and Sharma, 2015). However, being a long term investment in organic project and a higher density of small and marginal farmers in hills; organic farming faces a lot of constraint and limitations e.g.local market problems, less premium, more attack of pests and diseases, problems in certification. So, there is a need to construct infrastructure both technical and financial to motivate farmers to switch to organic farm practices.

• **Integrated Farming system** Sustainable development in agriculture must include integrated farming system (IFS) with efficient soil, water crop and pest management practices, which are environmentally friendly and cost effective (Walia and Kour, 2013)

It is the concept of judicious mixing of different enterprises like poultry, mushroom cultivation, fisheries, agro-forestry, goat/cow rearing and sericulture along with the main agricultural crop cultivation on a unit area which could help bring prosperity to farming. In hills, where resources are scare and opportunities for developing and adopting better technologies have lately declining the farm economic efficiency is an important factor to be considered productivity and growth. As conventional farming is becoming risky due to lot of problems, the farmers are reluctant to invest heavily in crop production so they want to shift to some promising enterprises thus IFS (integrated farming system) assumes greater importance for sound management of farm resources to enhance the farm productivity, reduce the environmental degradation, improve the quality of life of resource poor farmers and to maintained sustainability (Singh *et al.*, 2007). Farming system refers to the farm wherein two or more enterprises are integrated with that of farm resources for achieving the fuller utilization and realizing maximum profit and to stabilize returns. It is a resource management strategy that can achieve economic and sustained agricultural production to meet diverse requirements of the farm household while preserving the resource base and maintaining high environmental quality. The objectives of farming system in general are converging as to the development of sustainable location specific farm technology to raise and sustain the total farm productivity in terms of food ,feed, fodder and fuel (4F) to meet the felt needs of the farmers within the sphere of their agro-socio-political favourites and constraints. Farming systems aim for increased productivity, profitability, sustainability, balanced food, pollution free environment, recycling of unutilized resources, generation of income round the year, adoption of new technology, solving energy ,fuel and fodder crisis, avoiding deforestation, increased employment generation, literacy rates, input- output efficiency, enhanced opportunity for agriculture oriented industries and standard of living of the farmers. Generally farming systems practiced by farmers vary according geographical locations,climatological conditions,infrastructure facilities available,socially acceptable,socio-economic conditions of the farmers (Gill *et al*, 2007)

- **Farming systems in hilly and mountaneous areas:** integrated and holistic development of hills and mountaneous areas, need to be promoted by resource conservation techniques on watershed basis for improving productivity, profitability and thereby removing hunger and poverty. Due to the fragmentation of holdings, the per capita availability of land is declining day by day, no single farm enterprise is able to meet the growing demands of food and other necessities of the small and marginal farmers (Singh *et al.*, 2006). Integrated farming systems sustained crop productivity by reducing the risk of soil degradation, preserving soil's productive potential and decreasing the level of inputs required. Judicious mix of cropping systems with connected enterprises viz. fruits, vegetables, flowers, dairy, poultry, duckery, piggery, goatary, fishery sericulture etc. as suited to the specific agro-climatic conditions and socio-economic status of the farmers and shall be able to produce additional employment and income for the small and marginal farmers (Behera *et al.* 2004). Farming Systems" represent the integration of farm enterprises such as cropping systems, horticulture, animal husbandry, fishery, agro-forestry, apiary etc. for optimal utilization of farm resources bringing prosperity to the farmers.

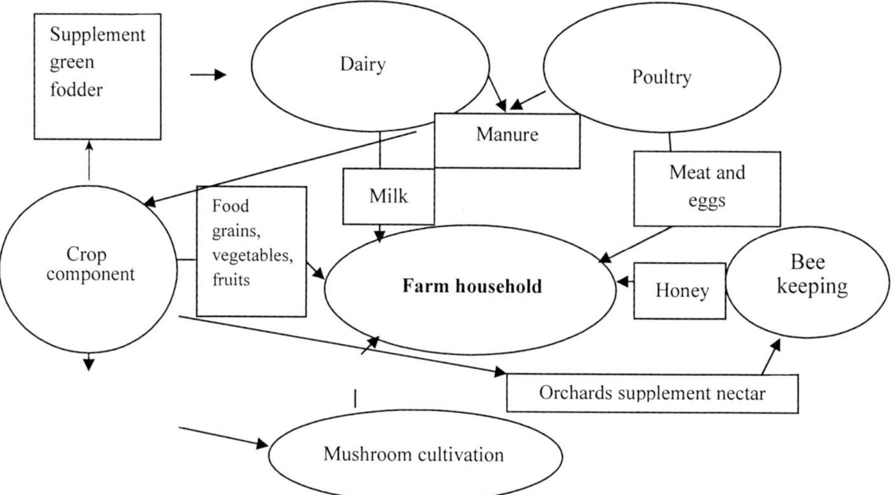

Fig. 2: The proposed Resource flow model of integrated farming system-hilly conditions (0.50ha)

Crop Diversification on Sloping Land The movement of soil, water and nutrients is the major problem on sloping land .Cultivation of crops in rows in sloppy lands causes soil erosion. In these type of fields, the crops particularly cereals, fodder crop etc. should be broadcasted and the plants remain haphazardly in field. As a result, the movement of water gets obstacle and more water is absorbed in the soil, thus reducing soil erosion. Mixed and intercropping (Cowpea-with cotton ,maize with rajmashetc.) practice checks the soil erosion and avoids the risks of the crop failure. Cropping systems should be designed in such a way that the soil is almost permanently covered with plant canopy. Leguminous plants such as *Sesbania cannabina, Crotalaria strata, Cassia tora, Vigna indica, Tephrosia candida, Leucaena glauca* and *Medicago hispida,* are used on sloping land either for soil loss reduction or green manure In arable crops, careful timing of sowing and planting can help to avoid uncovered soil being washed away during the rainy season. After the main crop is harvested, a green manure crop may be sown. On slopes, crops should be grown in contour lines across the slopes (along the contour lines) rather than vertically. Diversified cropping strategies such as mixed/ intercropping, strip cropping, alley cropping and agri-horticultural systems are developed to retain maximum amount of rainfall *in situ* and ensure higher production and protection against erosion (Gupta and gupta, 2010).

Intercropping in maize crop with rajmash

Horticulture. The hill areas have a comparative advantage in the production of horticultural products including growing of various horticultural crops like fruits, flowers and vegetables due to varied climate and thus making it an ideal location for growing temperate, subtropical and tropical fruits and off-season vegetables. Also, the production of fruits like apples further requires less water than vegetables and can be more successfully grown in areas that lack good irrigation facilities and may be more suitable in the relatively higher mountain ranges whereas, vegetables can be grown more successful in the foothills and valleys having irrigational facilities. The problems of inadequate information about best crops and best practices that are faced by farmers are important for the development of horticulture. The planning for horticultural development has to include supply of appropriate inputs for these best practices to the farmers, a scientific market analysis and the scientific methods of production and processing of horticulture products. (Azad *et al.*, 1988)

Apple farming

Agro forestry It is an intentional integration of agricultural and forestry based land –use-systems to provide tree and other crop products, and simultaneously it protects, conserves, diversifies, and sustain vital economic, environmental, human and natural resources. In agroforestry systems, there is an ecological and economical interactions among different components. As the hilly areas are encountered with problem of ever increasing demands of fuel, fodder, timber and non-tree products thus agro-forestry is important for meeting fodder, fuel wood and small timber of farmers, conserving soil and water, maintenance of soil fertility, controlling salinity and water logging, positive environment impact and alternate land use for marginal and degraded lands. Selection of proper land use systems conserve biophysical resources of non-arable land besides providing day-to-day needs of farmer and livestock within the farming system. For agroforestry system three types of systems are considered The first is a 'silvihorticultural' system where fuel trees dominant, the second type of system is an 'horti-agricultural' system where fruit trees are intercropped with crop plants and the third type of system is an 'agrisilvicultural' system where cereal crops and vegetables are grown as upland crops.Some medicinal and aromatic plants viz.ocium sanctum, Murraya koenogii, Azadirachta indica, Moringa oleifera, Mints, catharanthus roseus and coriandrum sativum can be commonly grown in Indian households. Jatropha curcas –an important source of bio-diesel, Pinus roxburghii is an important remunerative aromatic tree species, Acacia catechu having high medicinal value in addition to its value for fodder and timber can be grown in hills.the tree species like Azadiracta indica, terminalia spp. Aegler marmelos etc. can be intercropped with annual in early years until the tree canopy covered the ground (Gupta, 2010)

Olive in hills

Animal Husbandry In spite of the opportunities offered by favourable climate, the economic potential offered for dairying, sheep and other animal husbandry in the hill areas has not been fully tapped. In many hill regions the problem is of poor pastures and grasslands due to improper animal stock rate, non-rotational grazing, and poor management practices. The scientific management of these lands can increase the yield of fodder and support effectively a large animal population. The animal husbandry programme will need a strong preventive and curative animal health programme, together with processing and marketing of the produce The scientific management of these lands can increase the yield of fodder and support effectively a large animal population. The animal husbandry programme will need a strong preventive and curative animal health programme, together with processing and marketing of the produce (Anonymous, 2010)

Mechanization

As labour is getting scarce during peak periods , farm holding size and socio-economic background of hills is diverse and is mainly dominated by small farmers and poor farmers, the mechanization need to be focused on appropriate mechanization technologies. Mechanization program should not be limited to promote increased use of tractor but also include improved manual tools; animal drawn implements and appropriate mechanical machinery .As the majority of the farm operation is performed by women, the tools and machinery are to be promoted by catering the needs of women farmers. The research and development on farmer friendly appropriate tools and machinery is also needed to be reoriented (Manandhar *et al*, 2006)

Institutional interventions

Institutional interventions are needed to protect watersheds, stop annual cropping on lands beyond 15% slope and covering the lands having slopes 18-30% with forests under strict regulations, encouraging farmers to convert from inorganic farming to organic farming, It is evident from the land size distribution and economy of the hill agriculture that each household cannot purchase full set of **farm machinery** for its own use. These custom-hiring enterprises could play an important role on introduction of improved farm machinery and spread the benefit to the farmers in the community. Hence, government should recognize this custom hiring enterprise and provide them support through training and credit (Manandher *et al.*,2006)

Infrastructure and marketing

Hilly and mountaneous areas are characterized by severe infrastructure hurdles, especially with regard to connectivity of roads, transport network, electricity, irrigation, marketing infrastructure for agricultural produce, health , education and financial institutions (Bajracharaya,1992). There is an urgent need to address these problems problems with regard to connectivity with **road networks**, particularly of remote and inaccessible areas located in the mid and outer Himalayan ranges. This becomes more important for the marketing of perishable primary produce that include horticultural crops, which requires a well-developed network of roads connecting villages and agricultural areas to the urban areas and mandis. Also hills are gifted with a lasting source of water supply throughout the year. This means that there is tremendous scope for the development of small-scale hydro systems for **electrification** of the state, particularly in the far-flung areas of the Central Himalayan region marked by the absence of alternative sources of power , as ninety percent of the net sown area in hills is rainfed (Anonymous,2010). The hill areas have a comparative advantage in the production of horticultural products including fruits and vegetables. But due to lack of marketing facility, a large part of the total horticultural crops produced by hill communities is wasted each year thus there is a crucial need to develop an efficient marketing infrastructure in order to make it remunerative for the small mountain farmers to grow cash crops including fruits and vegetables (Azad,1988).

Efficient marketing system would consist of all post-harvest activities including the collection of farm products from the field, processing , packaging of the product, storing and warehousing of the product so as to address the problem of farmers having inadequate storage capacity by setting up of food processing units and cold chains and thus strengthening value addition. Also , there is a need to set up more bank branches at the district and local levels. Above all there is the problem of quality of education in hilly terrains. There is the need to increase public expenditure on **education** in order to improve the quality by promoting vocational and job-related education such as computer education. Access to health care in the rural parts of mountain districts continues to be poor. Moreover, there are issues related to **women's health** and nutrition, particularly in the remote hilly areas where access to basic health facilities is denied that need to be addressed. There is need to develop institutions that can offer cost effective solutions to problems of access and availability of health facilities for the rural mountainous regions. Hill women are the most important food producers and act as the invisible work force, thus less acknowledged backbone of the family economy. Even though, women fulfill a great number of essential tasks but they have limited access to and control over income, credit, land, education, training and information. It is only recently that **participation of women** in development programs in hilly areas is being considered necessary. The extension approaches and tools may still be gender biased and therefore much needs to be done to encourage cooperation and partnership of women in hill development.(Rijk,1989).

Tourism The main appeal for tourists in the hills and mountaneous terrains is the understanding of watching the snow capped peaks with singing rivulets, having clear blue sky, unparalleled majestic mountains, the vast expanses of alpine vegetations, diversified cultures, religious, adventurous wild tourism and much more, so,the tourism infrastructure must make it positive that this experience can be provided to the tourists without compromise to their comfort, and in new and innovative ways the development of tourism should be a planned approach maintaining the volume and quality of tourism, while trying to maximize the returns to the local economy – in terms of income and employment to the local people. In order to promote high-value tourism in the state, the sector has to facilitate a high quality tourism experience by making possible the infrastructure of hill stations, development of nature and adventure tourism with activities like skiing, paragliding, rafting, kayaking, etc., deploying high quality bus services encouraging car rental facilities, involving aerial tours of the hilly terrains using helicopters. availability of **quality hotels**. Overall, the tour operators and tourists agencies must be encouraged to corporatize and become more organized and professional (Driml and common,1995)

Policy interventions

To develop and flourish the growth of horticultural sector in the mountaneous states, all the problems listed above should have to be addressed by means of appropriate policies. It must also be clearly understood able that the overall policy package must try to solve all of problems simultaneously, as any one set of problems, if not addressed adequately, can significantly reduce the development of the sector. The **incentives** that the state can give to the small farmers is to shift to horticultural production that can also encourage the development of this sector. Thus the state needs to provide these farmers with subsidized inputs like seeds, fertilizers, insecticides, etc. Also, incentives should be such that to take care of the uncertainties due to market failure and natural calamities. The state should provide **minimum support prices (MSP)** in order to make sure that the poor farmers get good and remunerative prices. Finally, the provision has to be made to minimize farmer's losses due to natural calamites by means of **crop insurance**. For calamities that are relatively moderate in impact, small and marginal farmers must be supplied with poly-houses, poly-tunnels, hail nets, etc. However, for calamities that are very severe, the only protection for farmers can be through crop insurance schemes. The state should provide subsidies for the crop insurance of small and marginal farmers The policy should be to provide the farmers to sell their products with substitute options, so that the portion of the profit going to the middlemen can be minimized and the farmer gets a better price for his produce .Also farmers producing organic produce should be provided with organic certification so that produce can be marketed at a higher price. The small and marginal farmers need both scientific and financial help for their fightagainst natural calamities. **Latest appropriate farm machinery and implements** favorable policy is the critical for the promotion of agricultural mechanization in the mountain ranges and hills of the country addressing the needs of different category of farmers. All the different policy viz. co-operative farming and land consolidation for agricultural activity, contract hiring of agricultural machinery, support for agricultural machinery manufacturer, standardization and safety of agricultural machinery, promotion for energy efficient machinery, are needed to be streamlined for promotion of agricultural mechanization. **Tourism** there are some issues in tourism sector that should be addressed for example. discounts on standard prices' such as reduced airfares for particular groups, viz., students, senior citizens,disabled poersons and so on. The tourism policy must also break the seasonal pattern by developing winter tourism. **it** is equally important to make sure that the potential tourist has adequate information about the facilities available in the state. This will involve innovative campaigns through the media and the Internet that focuses on the factors that attract various types of tourism.

Conclusion

For developing right approaches to hill areas development all possible interventions through technology, policy , institutional etc should be framed. About 90% area of the hilly areas are rainfed , so watershed management technology should be introduced to secure the land and water conservation. New approaches introduced should meet the basic needs of hill people i.e. water, food, fodder, feed, fuel and employment. Major constraint in hills are marginal farmers with sloping and fragmented lands and to capture the advantages of available natural resources, integrated farming system with diversified agriculture provides sustainable productivity and profitability through resource recycling in all the microfarming. It will not only improve the marginalization of hill communities but also help in achieving social equity by building on the comparative advantages of key land resources. Due to isolation of hill communities, marginal farmers having limited access to agricultural technologies and inputs and thus breakdown of isolation by building infrastructure and opening up of hills to the wider market economy, tourism have great impacts on the livelihoods of small hill farmers.

5

Soil Conservation Measures

Soil conservation: it is using and managing land based on the capabilities of land itself, involving the application of the best practices to result in greatest profitable production without damaging the land

Principles of soil conservation

The aim of soil conservation is to obtain the maximum sustainable level of production from a given area of land whilst maintaining soil loss below a threshold level which theoretically, permits the natural rate of soil formation to keep pace with the rate of soil erosion. In addition, there may be a need to reduce erosion to control the loss of nutrients from agricultural land to prevent pollution of water bodies; to decrease rates of sedimentation in reservoirs, rivers, canals and ditches In the longer term, erosion has to be controlled to prevent land deteriorating in quality until it has to be abandoned and cannot be reclaimed, thereby limiting options for future landuse. Since erosion is a natural process it cannot be prevented, but can be reduced to an acceptable rate. A decision on what that rate must match the requirements for sustained agricultural production with those of minimising the environmental impacts of erosion.

Soil conservation measures

Broadly there are two methods of soil conservation. These are biological measures that includes agronomic measures, vegetative measures and mechanical or engineering measures

Agronomic soil conservation measures

Typically, agronomic measures are relatively cheap, requiring low inputs, can be very effective, and are often related to fertility management such as compost /manuring and thus to productivity. They are usually integrated into farming activities and often not considered as soil and water conservation (SWC) by the land users or specialists: these measures achieve conservation as a side-effect of good land management.

Agronomic measures have, in recent years, received much more attention. Perhaps the most notable example is conservation agriculture. Conservation agriculture rose to prominence when it was recognised by land users, rather than specialists, that with reduced tillage and lower costs erosion could be minimised, water used much more efficiently, and soil organic matter and biodiversity enhanced (Liniger and Critchley, 2007). There exist three major principles on conservation agriculture: minimal soil disturbance, permanent soil cover and crop rotations.

Land use based on its capability: Based on the capability or limitations the lands are grouped in to eight classes by the U.S. Soil conservation service :Class I to IV are used for agriculture or cultivation of crops,Class V to VIII are not capable of supporting cultivation of crops ,they should be used for growing grasses, forestry and supporting wild life

Contour farming: The carrying out of all agricultural operations namely, ploughing, harrowing, sowing, trenching, along contour lines is known as contour farming. Contour lines are lines that run across a (hill) slope such that the line stays at the same height and does not run uphill or downhill. As contour lines travel across a hill side, they will be close together on the steeper parts of the hill and further apart on the gentle parts of the slope

Even though the operation is very simple, it plays a major role in retarding the process of soil erosion as every furrow or crop row acts as a miniature reservoir to hold the excess runoff thus It conserves soil, and due to increased time of concentration, more rainwater seeps through the soil profile to recharge ground water. It has been reported that this operation increases crop yield by 10 per cent. Experiments show that contour farming alone can reduce soil erosion by as much as 50% on moderate slopes. However, for slopes steeper than 10%, other measures should be combined with contour farming to enhance its effectiveness

Dead Furrows: Dead furrows help in reducing the runoff velocity and thus conserve water. After the completition of all the tillage operations it is advisable to leave dead furrows at every 10 m interval. They should be remained in position until the crop is harvested.

Mulching: Mulches are ground covers, they prevent the soil from being washed away, reduce evaporation, increase infiltration, and control growth of unwanted weeds. Mulch can be organic crop residue, pebbles, or materials such as polythene sheets. Mulching prevents the formation of hard crust after each rain. Organic mulches add plant nutrients to soil upon decomposition. Use of blade harrows between rows also creates “dust mulch” by breaking the continuity of capillary tubes of soil moisture. Under indian conditions ,the economic aspect assume more importance.

Effect of mulches on crop yield

crop	Yield (q/ha)	
	No mulch	Mulch
Wheat	23.3	29.3
Barley	17.5	19.1
Sorghum	5.3	9.4
Tobacco	13.3	18.4

Source Randhawa and Ventaswarlu ,1980

Cultivation of proper crops: Cultivation of row crop in sloppy lands permits the soil erosion. In this filed, the crops particularly cereals, fodder crop etc. should be broadcasted and the plants remain haphazardly in field. As a result, the movement of water gets blockage and more water is absorbed in the soil, thus reducing soil erosion. Mixed and intercropping (Cowpea-with cotton ,maize with rajmashetc.) practice checks the soil erosion and avoids the risks of the crop failure.

Cropping systems should be designed in such a way that the soil is almost permanently covered with plant canopy. In arable crops, careful timing of sowing and planting can help to avoid uncovered soil being washed away during the rainy season. After the main crop is harvested, a green manure crop may be sown. On slopes, it is good if crops should be grown in contour lines across the slopes (along the contour lines) rather than vertically

Table Run off and soil loss as influencd by different cropping system

Cropping system	Runoff (mm/year)	Soil loss (t/ha/year)
Castor	98	2.89
Sorghum	81	2.31
Sorghum+redgram	76	2.21
Pearl millet	65	1.38
Anjana grass	42	0.37
Cultivated fallow	130	4.71

Source: Reddy and Reddi (2005)

Tillage practices used for soil conservation

Tillage practice	Description	How it help
Conventional	Standard practice of ploughing with disc or mould board plough,one or more disc harrowing,a spike tooth harrowing, and surface planking	Rough cloddy surface,intercept more water and less run off.

No tillage	Soil undisturbed prior to planting which take place in a narrow,2.5-7.5cm wide seed bed.crop residue cover 40-100 per cent if maintained on surface,weed control by herbicides	Not suitable for soils which are compact and seal easily Surface runoff reduce because of mulch water stable aggregates more as compare to traditional ploughing
Strip Tillage	Only isolated bands of land are tilled with the intervening areas left undisturbed. weed control by herbicides and cultivation	Causes less compaction,conserve soil moisture and reduce losses of organic matter, nitrogen, phosphorus and potassium
Mulch tillage	Leaves the residues of the previous crops atleast 30 % on the surface of the field. weed control by herbicides and cultivation	Control both water and wind erosion and conserve moisture in the drier areas. Physical conditions of soil improves as no direct splash,less dispersion of surface soil, less fluctuation in soil moisture and temperature, increased WHC, increased amount of percolation
Minimum Tillage	Any other tillage practice which retains at least 30% residue cover	Improve soil condition due to insitu decomposition of plant residues, higher infilteration caused by vegetation present on the surface

Source (Gupta and Gupta, 2010)

Mixed and inter cropping: Mixed cropping of different crops along with the main crops, such as millets and different legumes, is better and continued cover on land .The different root systems of mixed crop feed at different depths of the soil, thus there in no fear of soil fertility depletion.

Relay cropping in this different crops are grown in a rotation but here the next crop is planted before the previous crop is harvested thus provide a partial cover to the soil and hence control erosion.

Crop rotation: It is a planned sequence of cropping. crop rotation is an important method for checking erosion and maintaining productivity of soil. In good rotation densely planted small grain crops, spreading legume crop etc. should be included which may check soil erosion.

Use of fertilisers and organic manure: Fertilisation of crop is not only important in the nutrition point of view but it also help in the better development of root and shoot system and thus conserve soil moisture. Adding organic manures such as farmyard manure and compost every year as basal application to the soil improves the physical condition of the soil considerably. The crumb and granular structure increases the infiltration and permeability in the soil and conserve the soil water. Consequently soil erosion decreased.

Plant population: Use of recorded seed rate /plant population provides opportunity of compete surface coverage and thus reduce the soil loss Normal season : Sowing is done with the normal seed rate. some crops make up the population due to tillering like wheat barley rice but crops like maize ,sorghum etc. do not make density later on .therefore it is of utmost important to maintain proper population.Late and drought season : Where the monsoon is moderately delayed, normal cropping with reduced seed rate is advised. However, if there is a drought during the plant's growth period and wilting is likely to occur, selective thinning is recommended to reduce the plant population to effectively use the scarce soil moisture among fewer plants.

Line-Sowing: Line-sowing on contours is essential. It arrests runoff and conserves soil being eroded. It helps in the use of labour-efficient implements in weeding (i.e., removal of unwanted vegetation through use of different sizes of blade harrows between the rows).

Varietal variations: Varietal requirements depends on the agro climatic conditions of the soil and local needs apart from being high yielding, the improved varieties have many desireable characters like plant type,disease resistance, responsive ness to fertiliser and irrigation vis –a vis hardness to moisture stress conditions.

Where the distribution and the amount of rain is unpredictable, it is important to select varieties which have a shorter duration life cycle (seed to seed) to cut down the water requirements of the crop. In drought-prone areas, the success rate of short duration crops is greater than long-duration crops.It is important that varieties which have proven genetic character to withstand longer periods of drought are chosen so that the crops can do well even in situations where the intervals between rainy days are long.

Weed and pest management: Uncontrolled weeds remove soil moisture and nutrients otherwise meant for the crop. Yield losses varying from 10-90% in different crops depending upon weed flora and density so frequent weeding is an important part of agriculture. Management of weeds save fields from being sick due to obnoxious weeds. Line sowing and mechanical weeding, with appropriate size of blade harrows, remove unwanted vegetation which competes with the main crop. Protection of crops from various diseases and insect-pests also help in saving the crop

Vegetative soil conservation measures

Strip Cropping: It consists of growing erosion permitting crop (e.g. Jowar, Bajra, Maize etc.) in alternate strips with erosion checking close growing crops (e.g. grasses, pulses etc.). Suitable width of erosion permitting and erosion resisting crop have to be adopted on different slopes. In general, 3:1 ratio of erosion permitting and erosion resisting strips.

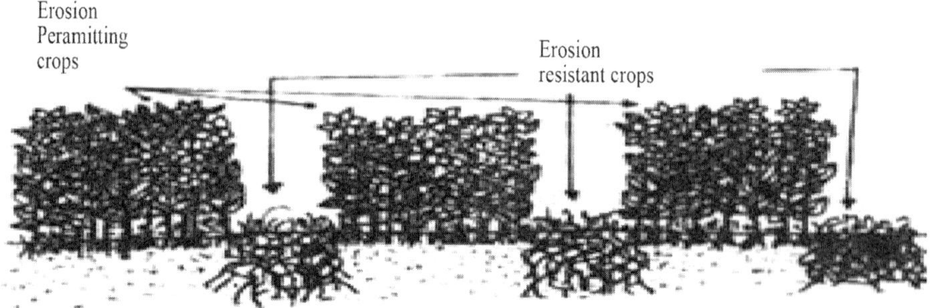

It may be of different types as follows:

Contour strip Cropping: Contour strip cropping is the growing of erosion permitting and erosion resisting crops alternately in strips across the slope and on the contour line. This practice is useful because it checks the fast flow of run-off water increases the infiltration of water in the soil and prevents soil erosion.

Field Strip Cropping: Strips of crop are parallel to the general slope of the land.

Wind Strip Cropping: Strips of crop are across the direction of wind regardless of contour.

Buffer Strip Cropping: In this, the severally eroded portion of land is permanently kept under grass and contour strip cropping is practices in the rest of the area.

Cover crops: Keeping the soil covered is a fundamental principle of conservation agriculture. Cover crops are beneficial as they:

- Stabilize soil moisture and temperature
- Protect the soil during fallow periods
- Mobilize and recycle nutrients
- Improve the soil structure and break compacted layers and hard pans
- Permit a rotation in a monoculture
- Can be used to control weeds and pests
- Produce additional soil organic matter and improve soil structure

The grasses and legumes together produce very dense sod that helps in reducing soil erosion. Sod forming crop e.g lucern (*Medicago sativa* L), Egyptian Clover or Berseem (*Trifolium alexandrinum*), ground nut (*Arachis hypogea* L), Sannhemp (*Crotolaria juncea*), grass etc. have ability to cover the surface of the land and their roots and further bind the soil particles to form soil aggregates, thus help in preventing soil erosion.

Afforestation: Afforestation means replenishing of forests where there were no trees it may be due to adverse factors such as unstable soil, aridity or swampiness. Along with afforestation, reforestation should be undertaken that means replanting of forests at places where they have been destroyed by uncontrolled forest fires, excessive felling and lopping. Afforestation is the best means to check the soil erosion. Lutz and Chandler (1946) cited the following points in support of vegetational check erosion as follows:

- Infiltration of water is favoured due to high porosity of soil under vegetation. Percolation of water helps in preventing the soil moisture which accelerates further growth of the vegetation.
- Surface accumulation of organic matter increases the water holding capacity of the underground soil.
- Root system of vegetation holds the soil mechanically and provides stability of the underground soil.
- It gives the protection against wind. The forest vegetation shields the soil from direct effect of drought, snow and rain.

Control of grazing: Grazing can increases the soil erosion. But the grazing cannot be completely paused in all areas. So the restricted and rotational grazing may be promoted to check soil erosion to some extent. The area open to grazing for sometimes should be closed for the following year to facilitate regeneration of forests and to maintain thick ground vegetation

Agro-horticulture: Good annual crop returns are not obtained from the marginal lands even in normal season. But these kinds of soils are best used for raising trees of economic value and creating permanent assets. Some of these lands are also very good for raising dryland horticultural crops such as mango, ber, jamun pomegranate, tamarind etc. A part of the land could be marked specially for planting mixed tree species known in the area for providing fuel, fodder and timber for household needs and agricultural implements. Trees provide stable and sustained income every year, especially in drought years.

Wind-breaks: Wind Erosion is as important as water erosion, the same system of cropping is recommended to be adopted, generally across the wind direction. Farmers can also raise wind belts across the wind direction. Alley cropping with rows of trees at 30 to 45m intervals also controls the desiccating effects of hot and dry winds.

Mechanical measures

Now a days soil erosion **has been looked upon as a** physical problem so there set a trend to treat erosion as an engeneering problem.some mechanical and engineering measures are **Contour bunds** these are the physical barriers that are built along the contours in series for diversing excess runoff safely and prevent soil from erosion. The eroded silt start depositing along the bunds and bunds take the shape of a riser. further these risers should be planted with the grasses to check erosion. These are recommended where rainfall is less and slope is less than 6% **Graded bunds** these are the small earthen bunds with the slight grade .they for taking runoff safely .these are recommended up to 10% slope but their efficiency gets lessened beyond 4% slope. They are helpful in reducing soil erosion and also insitu rain water conservation are constructed across the slope. **Bench terraces** these are the flat beds that are constructed on the hills across the slope to promote uniform distribution of runoff and control soil erosion **Half moon terraces** these are the semi-circular beds with shape that resembles a half-moon.these trees are for fruit trees and other plantations crops on steep slopes (Sharma,2010) **Surface ridging it** includes cultivating the soil to produce rough ,cloddy surface and convert soils in to ridges in lines across the direction of the prevailing wind to control erosion. **Stone lines it is** the oldest soil conservation practices best used in the mountaneous parts where stone lines or stone walls are built across the slope of fields on or close to the contour. The stone barriers slow down the runoff and thus stop moving soil. **Dams** they slow down runoff andcatch silt and thus control erosion. **Gully control** measures are constructed generally with the locally available materials. vegetationlike grasses,s hrubs and trees is one of the cheapest source for controlling gullies. Concrete and stone structure are used at the gullies top to stop them becoming bigger and silt traps can stabilize eroded gullies. **Basin listing** in this with the help of implement called basin lister small basins are formed along the contours to hold and prevent runoff **Sub-soiling** in this with the help of sub-soiler the soil is broken in to fine grains to increase absorption capacity of soil .**Contour terracing** these are made to check the runoff,in these series of ridges or bunds of mud are formed along the contours they further includes **Channel terrace** where a shallow channel is dug out and silt is deposited along the lower edge of the canal. **Broad base ridge terrace** in these wide canal is made on the contour by excavating the mud. **Narrow based ridge terrace** if canal is narrow never wide than it is called narrow based ridge terrace. **Bench terracing** in it series of platforms are made along the contours across the general slope of the land. **Contour trenching** in this across the slopes trenches of two feet by one feet are formed and these trenches are used for planting tree seedlings.**Terrace outlet** these are structures which are constructed for safe disposal of the run water **Ponds** these are the water harvesting structures which are made at suitable places to store runoff. **Stream bank protection** in this concrete or stones protective walls are constructed on banks of channels or rivers to prevent stream bank erosion. **Diversion drains** these are the constructions made in arable area and non-arable area to divert runoff which is coming from the slopy lands to the safer places and should have an adequate outlet.they should be constructed with a non-erosive or non-silting velocity **Water ways** they are made in the conservation system with the main purpose to convey the runoff to a suitable and safer outlet they are of three types **diversion channel** these channels are built at a slight grade on the upslope areas to intercept water from the top of the hillside to drain runoff safely to the outlet **terrace channe**l these channels are also built across the slopeto collect runoff from inter-terraced area to a suitable outlet, **grass water ways** these are located on natural depression of the hill side. these channels are most effective in moderating the flow and reducing velocity of run off with the suitable grasses planted on the runoff route.the shape of the waterways depend upon the field conditionsand type of equipment used for the construction.generallty parabolic,triangular and trapezoidal shapes are used for construction.

Soil management practices

Maintaining soil pH soil's acidity or alkalinity is measured by the soil pH levels. Addition of basic or acidic pollutants made soil to get polluted and it can be correcting by maintaining the desirable pH of soil. Acid soil is managed by liming, it inactivates Al, Fe and Mn thus lessened fixation of P increase microbial activity, enhances uptake and thus productivity. Wheras alkalinity is managed by using gypsum **Salinity management as** metabolism of the plants is harmed by the excessive collection of salts in the soil it can lead to death of the vegetation and thus cause soil erosion, so salinity management is important. These soils are reclaimed by leaching of excess soluble salts with irrigation water having good quality, lowering water table by draining excess water, growing salt tolerant plants and varieties, using humus in the soil and involving good management practices **Soil organisms** these are the important component of the soil, they make nutrients available for plant uptake. The beneficial soil organisms like earthworms increase porosity of the soil. research shows that earthworm casts are five times richer in available nitrogen, seven times richer in available phosphates and eleven times richer in available potash than the surrounding upper150 mm of soil. The weight of casts produced may be greater than 4.5 kg per worm per year. By burrowing, the earthworm is of value in creating soil porosity, creating channels enhancing the processes of aeration and drainage.(Mollison,1988). Biofertilisers contain micro-organisms so the inoculation of these fertilizers increase yield and also increase the availability of major nutrients like N,P,K and S (Faroda *et al.*, 2007)

Contingent Planning

With every care taken to undertake timely agricultural operations, it is still possible that the whole operation becomes a gamble due to unpredictable monsoons. The main crop could fall in the early part of its life cycle. In such cases, the farmer should come up with an alternate crop that can mature in a very short time and under hard conditions to take advantage of what is left of the rainy season. Contingent planning helps catch and make the best use of late rains. Advance planning is necessary in selecting a contingent crop. And all the requisites for its sowing should be ready within the main season itself. Credit for farmers must be made available at the right time.

Conclusion

Success of soil conservation schemes depend on the nature of the erosion problem identified, suitability of the conservation measures selected to deal with the problem. A sound land use plan is important for soil conservation and what it is best suited under present or proposed economical and social conditions. lan

By adopting the land capability classification as the methodology for landuse planning, the distinction will be made between areas where erosion is likely to occur when the land is used in accordance with its capability and that which will arise from misuse of the land. Once the most appropriate landuse has been determined, soil conservation is a matter of good management of the land. Erosion-control measures proposed must be relevant to the farming system.

6

Moisture Conservation for Mitigating Livelihood Problems of Rainfed Areas

Introduction

For sustainability hill agriculture water conservation issues are of great concern. Water is a generally limiting factor for crop production in hills and further water scarcity is arising at alarming rate due to climate change, declining soil productivity and inefficient management factors. Global warming is also putting the serious impact on the intensity, frequency and distribution of the rainfall. Further hill soils are ecologically fragile with undulated steep soils with low water retentive capacity where irrigation is not available. Moreover scope of supplying water by developing new water resources is limited thus under these conditions there is an urgent need for development of recent advanced agrotechniques to conserve and harvest natural water resources and protect the fragile environment of hills.

In situ moisture conservation and retension-soil measures

Crop yield in rainfed areas are very low and unstable. Moreover due to impact of climate change there is occurring spatial and temporal variability in rainfall that cause severe gaps in water flow and ground recharge thus following in situ moisture conservation techniques are essential for these areas.

Composting is the microbial decomposition of usually local resources such as crop residues, household refuses and animal manures that produce high quality organic decomposed material. The simplest way to produce compost is the use of simple heaps where the material is accumulated and the natural decomposition processes are favoured by maintaining an optimal humidity and temperature and by ocassionaly turning over the decomposing matter (Hudson,1987).The use of compost is of particular importance in poor soils having low percentage of clay and where water lost through percolation (Quedraogo,2001) as they have high water holding capacity and that improves infilteration.

Mulching it is organic or inorganic material of synthetic or natural origin applied on soil surface to check the evaporation and soil water. About 70% rainfall is lost through evaporation which can be reduced by mulching. When mulch is applied the non-productive use of green water is diminished causing an augmented availability of plant uptake and therefore, an improved water productivity (Karlberg *et al*,2009) **.** The most common low cost material used in agriculture are agricultural residues, cut weeds and straws (Deveskog, 2001).other materials used for mulching are polythene sheets , rocks etc. The positive outcomes of mulch farming are reducing the impact of the rain drops, decreasing flow velocity by providing roughness to the soil surface, and thus improving infiltration capacity. It also increase the burrowing activity of some species of earthworms (e.g. *Hyperiodrilus* spp. and *Eudrilus* spp. which indirectly improves transmission of water through the soil profile and reduces surface crusting and runoff and improves soil moisture storage in the root zone. The amount of mulch required for structural stability and infiltration capacity is dependent on the rate of residue decomposition, type of climate, soil properties, relief and amount of rainfall '. Studies relating soil loss to bare ground indicate that about 70% of the soil surface must be covered by mulch to be effective The application of mulch @ 8tons/ha on the soil surface proved to be an effective method to reduce runoff and associated nutrient losses((Khera and Hadda,2010). **Soil Conditioners** are oil or rubber based emulsions of poly-functional polymers that make aggregates in soil by making chemical bonds with the clay minerals and thus increased infilteration rate by improving size and stability of the pore spaces . Among the several conditioners used for water runoff control bitumen treatments are effective to control storms, for high infiltration rates Poly-acrylamide conditioners regardless of the size or distribution of aggregates may be used, whereas among the emulsions asphalt and latex may be used, for the best results soil conditioners should make aggregates which must be at least 2 mm in size and ideally greater than 5 mm.

Soil carbon sequestration it is a good solution for sustaining soil health in both irrigated and rainfed areas . Organic matter added by FYM, compost, green manures, waste plant residues, farm weeds or even crop residues supplies essential nutrients, food for microbes and improved soil properties. Soil organic carbon also influences the water holding capacity and available water capacity of soils.studies have shown that 1% increase in soil organic matter increases field capacity by about 2.2%,permanent wilting point about 1% and available water capacity by 1.2%.

Table: Effect of land modifying measures in water erosion

Types of water erosion	Untreated	Treated
Sheet	80.0	15.0
Rill	11.0	2.0
Gully	3.6	Nil

Source:(Khera and Hadda,2010)

***In situ* moisture conservation and retension-Agronomic measures**

Conventional tillage are those that leaves less than 15%of crop residues on the soil surface and these operations increase porosity and help in root development and infiltration thus minimise runoff by assisting infiltration. Sucess of ploughing in soil and water conservation depends on the conditions under which it is carried out and its frequency. One thing that is taken in to consideration during ploughing is that soil moisture content are right to avoid structural breakdown and smearing, **Special tillage operations** viz. soil inversion, chiselling, subsoiling and deep tillage (for soils with an impeding layer within rooting depth) are proved to be very beneficial in minimising soil hardening, improving surface detention, storage, infiltration, and root development. **Conservation tillage** are those types of tillage that leaves at least 30% of crop residues on the soil surface . It is very important to minimise the pulverizing effect of conventional tillage. It can be **strip zone tillage** in this the small strips of land required for seedbeds are only cultivated thus leaving wide untilled zone **. Mulch tillage** in this type the tillage is carried out by retaining the mulch in the ground**. Minimum tillage** in this least tillage activities are completed with plough-plant operations as possible in one pass**. Zero** tillage in this tillage soils remain untilled for many years which left crop residues on the soil surface and thus produce the layer of mulch, which protects the soil from physical impact of the rain, stabilises the soil moisture, temperature favouring the habitats of number of the organisms viz. larger insects, soil borne fungi and bacteria. Further these organisms chew up the mulch, decompose it enrich soil with the organic matter which provides buffer for water and the nutrients. Where as, larger components of the soil fauna such as earthworms produce stable soil aggregates as well as uninterrupted macro-pores allowing fast infiltration in the heavy rainfall (Shukla, *et al*,2011) From the studies it has been observed that as much as a 5-fold reduction in runoff has been reported under no-tillage compared to conventional tillage **. Reduced tillage** refers to those type of the tillage techniques that leaves crop residues less than 30% and is very important to optimise the use of green water. Contour farming it is the carrying out of all agricultural operations namely, ploughing, harrowing, sowing, trenching, along the contour lines this farming system is reffered as contour farming. It involves aligning plant rows and tillage lines at right angles to normal flow of runoff. Even though the operation is very simple, it plays a major role in retarding the process of soil erosion as every furrow or crop row acts as a miniature reservoir to hold the excess runoff thus it conserves soil, and due to increased time of concentration, more rainwater seeps through the soil profile to recharge ground water. It has been reported that this operation increases crop yield by 10 per cent. Dead Furrows Dead furrows help in reducing the runoff velocity and thus conserve water. After the completion of all the tillage operations it is advisable to leave dead furrows at every 10 m interval. They should be remained in position until the crop is harvested. **Ridge and Mound Tillage** ridge-furrow system used along the contour will enhanced infiltration and reduced runoff. These include: terraces and contour bunds to break up long slopes; and diversion channels and watercourses to dispose safely inevitable runoff. **Terrace Farming** it involves earthworks made at right angles to the steepest slope consisting of an excavated channel on the uphill side, the spoil from which forms a bank on the downhill side. Terraces are made by different techniques and according to method of construction they are called bench terraces; broad-based and narrow-based types; Nichols terraces; diversion terraces with mangum; channel terraces; retention terraces etc. these terraces are very helpful in converting a steep slope into a ladder with horizontal ledges and vertical walls of stone, brick or timber between ledges. Retention and bench terraces are generally usable on slopes up to 8% and slope of 12% to 40% respectively. Along with different tillage system green water use is also optimise by adopting judicious farming systems.

Multi-storey cropping: It is the cultivation of different plant species with different height and growth characteristics that use soil, nutrient and moisture from different layers without creating competition (Liniger and Critchley, 2007).

Alley cropping provides an alternative farming system by integrating perennial trees and shrubs with annual food crops. To provide *in situ* green manure and to prevent shading of crops arable crops that are grown in the spaces between rows of planted woody shrubs or trees are pruned during the cropping season. Alley-cropping arable crops (such as maize) grown with *Gliricidia* and *Leucaena give the* promising results. The permanent rows built of trees along the contours decrease runoff and favour the infilteration.

Strip Cropping: This is farming of sloping land in alternate contoured strips of intertilled row crops and close-growing crops (for example, a cover crop or grass) aligned at right angles to the direction of natural flow of runoff. The close-growing strip slows down runoff and filters out soil washed from the land in the intertilled crop. Usually, the close-growing and intertilled crops are planted in rotation. Strip cropping provides effective erosion control against runoff on well-drained erodible soils on 6 to 15% slopes. The width of the strips is varied with the erodibility of the soil, and slope steepness. Strips along the contours can be left uncultivated with an evergreen leafage of natural plants with the function of slowing down runoff and increasing filteration (Liniger and Critchley, 2007). **Cover crops** acts as in-situ mulch increase soil organic matter and thus helps in conserving the soil water and infiltration capacity .

The effectiveness of cover crops depends on the type and characteristics of species viz. ease of establishment, vigour of growth, depth of rooting, rapidity of establishment of surface cover etc. Cultivation of proper crops cultivation of the row crops in sloppy terrains without taking precautions can permits soil erosion. In this filed, the crops particularly cereals, fodder crop etc. should be broadcasted and the plants remain haphazardly in field. As a result, the movement of water gets obstacle and more water is absorbed in the soil, thus reducing runoff. **Mixed and intercropping (**Cowpea-with cotton, maize with rajmash etc.) practice checks the soil erosion and can avoids the risks of the crop failure. Cropping systems should be designed in a way that the soil is almost permanently covered with plant canopy architecture . In arable crops, careful timing of sowing and planting can help to avoid uncovered soil being washed away during the rainy season. After the main crop is harvested, a green manure crop may be sown and grown . On slopes, crops should be grown in contour lines across the slopes (along the contour lines) rather than vertically

Table: Run off and soil loss as influencd by different cropping system

Cropping system	Runoff (mm/year)	Soil loss (t/ha/year)
Castor	98	2.89
Sorghum	81	2.31
Sorghum+redgram	76	2.21
Pearl millet	65	1.38
Anjana grass	42	0.37
Cultivated fallow	130	4.71

Source:Reddy and Reddi (2005)

Good use of agrotechniques also favour the infiltration rate ,and increase water productivity

- **Plant population:** Use of recorded seed rate /plant population provides opportunity of compete surface coverage and thus reduce the soil loss
- **Normal season** : Sowing is done with the normal seed rate. some crops make up the population due to tillering like wheat barley rice but crops like maize ,sorghum etc. do not make density later on .therefore it is of utmost important to maintain proper population.
- **Late and drought season**: Where the monsoon is moderately delayed, normal cropping with reduced seed rate is advised. However, if there is a drought during the plant's growth period and wilting is likely to occur, selective thinning is recommended to reduce the plant population to effectively use the scarce soil moisture among fewer plants.

- **Varietal variations:** Varietal requirements depends on the agro climatic conditions of the soil and local needs apart from being high yielding, the improved varieties have many desireable characters like plant type,disease resistance ,responsive ness to fertiliser and irrigation vis –a vis hardness to moisture stress conditions. Where the distribution and the amount of rain is unpredictable, it is important to select varieties which have a **shorter duration life cycle** (seed to seed) to cut down the water requirements of the crop. In drought-prone areas, the success rate of short duration crops is greater than long-duration crops. It is important that varieties which have proven genetic character to withstand longer periods of drought are chosen so that the crops can do well even in situations where the intervals between rainy days are long.

- **Grafting:** Vegetatitive propagation techniques like grafting with a robust wild root stock is combined with an aerial part of another cultivar of same species but with more performing characteristics improve resistant of trees to drought.

- **Weed and pest management:** Uncontrolled weeds remove soil moisture and nutrients otherwise meant for the crop. Yield losses varying from 10-90% in different crops depending upon weed flora and density so frequent weeding is an important part of agriculture. Management of weeds save fields from being sick due to obnoxious weeds. Line sowing and mechanical weeding, with appropriate size of blade harrows, remove unwanted vegetation which competes with the main crop. Protection of crops from various diseases and insect-pests also help in saving the water.

- **Use of fertilisers and organic manure:** Fertilisation of crop is not only important in the nutrition point of view but it also help in the better development of root and shoot system and thus conserve soil moisture. Adding organic manures such as farmyard manure and compost every year as basal application to the soil improves the physical condition of the soil considerably. The crumb and granular structure increases the infiltration and permeability in the soil and conserve the soil water.

- **In-field water retension:** Techique refer to the structures that allow soil in a cropping field to retain or store moisture for a specific period enough to cater for a dry spell. They also provide additional sevices like fodder , fertility improvement and fuel (Ngigi,2003) in hill farming large amount of runoff is often generated produce erosive features like rills and gullies. Bunds, ridges and terraces are the simple and the common techniques which increase infilteration and reduces runoff. **Contour bunds** can be created with common materials like stones, dug soil or with the remains of the crops. Where the land usually has a gentle, long and continuous slope, stone lines laid along the fields with a spacing of 15-30 meteres in between lines, slow down runoff ,spread it evenly and favoured cultivation (Critchley and Reji,1989). **Earth bunds** with a spacing of 5-10m and with upward ties are used for cultivation of trees and crops in the inter-row spacing (Critchley, 1991). Earthen ridges are usually smaller and used for crop cultivation in the zone immediately adjacent to the upper side of the line where a shallow furrow has been dug to build the ridge (Critchley *et al*,1991). Weeds and crop residues are laid in bands across the slope of annual crop fields to conserve soil, water and to incorporate organic matter in to the soil after decomposition(Liniger and Critchley, 2007).Tied ridging are also a series of basins created in the field in order to retain water by retarding runoff (Georgis *et al*, 2001).

Water harvesting

It is the process of collecting, concentrating and improving the productive use of rain water and other forms of precipitation, and reducing unproductive depletion (Dhar,2010). Water harvest techniques that is collected from the hill slopes and man made catchment are the simplest and cheapest based on indigenous and the traditional system. It has been observed that 1mm of rainfall can produce 1000 litres of water per hectares and hence small impermeable area can produce a relatively large volume of water (Gupta and Sharma,2010). As in hills the scope of the installation of irrigation projects are meagre due to constraints of suitable sites, funds, equity and other social factors so the water harvesting techniques can play a major role in those areas. It has been found that rain water harvesting is also possible in those areas having 50-80mm average annual rainfall but often the catchments need alterations and modifications so that soil surface becomes impermeable and generate more runoff. **Land alteration** in this method the runoff concentration is increased by clearing away rocks and vegetations and further compacting the soil surface by giving it **chemical treatments.** Chemicals like sodium fills the pores and others chemicals like ,silicon latexes, asphalt and wax act as water repellent and act as a soil sealent. **Soil covers** like plastic sheet, butyl rubber, metal foils and gravels can also help in building the low cost catchments and inducing runoff.

Source of water which is to be harvest may be from surface, roof, runoff, in-situ and flood water harvesting. **Surface water harvesting** allow conservation of moisture in the field by collecting surface runoff thereby recharging groundwater, and as a source of irrigation for the crops (Subgyono,K.2007). This technique may be short term and long term. Short term water harvesting is designed to collect the 24-hour rainfall excess for a ten year recurrence interval (Gupta and Sharma,2010). In this type water is harvested in field bunds, contour bunds, ditches and trenches etc. whereas, in long term, water harvested is used for the purpose of agriculture, irrigation dinking, fishing, recharging ground water etc. This type of technique is completed by constructed a reservoir in an area and most common type of measures are farm ponds, percolation ponds, sunken ponds etc. **Roof water harvesting** this type of harvesting is mainly applied for the domestic use, in drinking by humans or by livestocks as it is free from the soil pollutants. **Run off water harvesting** it includes mainly runoff collection and storage techniques ,in this technique rain water runoff flows naturally towards the water reservoir by gravity (Nissen,2006). These techniques besides providing irrigation facilities also control sediments erosion and act as a flood control mechanism. It may includes on-site and off-site runoff collection. In on-site runoff collections the reservoirs are mainly tanks and cisterns and the water stored in them are usually used for the domestic purposes and watering vaegetable gardens and fruit trees on the home stead (Subgyona and Pawiton, 2008). Whereas, in the off site run-off collection the storage structures are usually open including dams, pools etc. the water stored in these reservoirs are mainly used for the livestock and the agricultural purposes. **Run off diversions** in run off water harvesting water is harvested from the streets, roads either paved or unpaved, these diversions acting as a rainwater catchment area generates runoff and these runoff being channelised in to the fields or in the storage structures for multiple purposes. **In-situ rain water harvesting in** this type of water harvesting several options are opted, for impounding rainwater land levelling and field bunding can be done. Sowing of crops in the furrows can also play a great role in harvesting the rainwater and enhances the crop yield . futher the technique of sowing of high water requiring crops in sunken beds and low water requiring crops on raised beds meet the requirement of the different crops in hills. **Flood water harvesting** the basic principle in this category is that water is collected secondarily for the agricultural purposes whilst the primary purposes are for controlling concentrated water flow during or after a heavy rainfall .Techniques in this category allow controlling of soil erosion, collection of fertile sediments, conserving runoff, recharging aquifier, moisture conservation and harnessing underground water flow in river beds for later extraction (Mengistie,1997).

Water harvesting structures at the foot hills slopes and hilly areas rain water may be harvested in the small or large water harvesting structures with a water storage capacity for an area of about 0.4 to 0.5 ha **Check dams are the** temporary or permanent structures . Temporary check dams are constructed with locally available materials like brush wood, loose rocks and woven wires and are helpful in impeding the soil and water erosion. These structures are cheap, and can lasts about 2-5 years. Whereas, permanent check dam are constructed with stones, bricks and cement.

Percolation Pond: These are the multipurpose conservation structures that stores water for livestock and also recharges the groundwater. This structure is constructed by excavating a depression, forming a small reservoir or by constructing an embankment in a natural ravine or gully to form an impounded type of reservoir. If water in these structures are used for irrigation, it is sufficient to irrigate 4-6 hectares of irrigated dry crops (maize, cotton, pulse, etc.) and 2-3 hectares of paddy crop.

Farm Ponds: These are the small reservoirs constructed for the purpose of storing water essentially from the surface runoff. These may be of four types viz. dugout ponds, Embakement type ponds, spring fed ponds and off-stream storage ponds(Gupta and Sharma,2010)

Dug out farm ponds These ponds are generally constructed in areas having flat topography and are excavated at a particular site and the soil obtained from the excavation is used as an embankment around the pond.

Embakement type ponds: These are the surface ponds and are generally constructed in a site having natural depression .They are partially excavated with an embankment constructed on the down slope to retain the water. In these structuctures it is desireable to take the water out of the pond through a gravity outlet for the irrigation of the down field areas.

Spring fed ponds: These structures are made where spring or creek is the source of water supply to the ponds. These ponds are generally found in the hilly areas where natural springs or creeks are available. Runoff water from natural creeks is collected in ponds of the desireable size placed at appropriate sites.

Off stream storage ponds: It is a storage reservoir that is filled by diverting water in to it or pumping from other sources .These ponds are constructed by the sides of ephemeral streams which flow seasonally.

Irrigation Tank: These are the constructions which are made up of earthen bunds and are reinforced with masonry that collect and store rainwater for irrigation. In Tamil Nadu, India, each tank irrigates from 10 to 5 000 hectares. In south India, there are about 2 000 000 tanks, irrigating about 3.5 million hectares.

Fog water harvesting in the temperate hills the harvesting of the fog is worth mentioning. this is very important technique to provide water to the communities who suffer fro the water shortages. The system has already been successfully used in the South-America (Olivier,2002).

These storage facilities besides controlling the floods during storms will also enable proper moisture conservation and also facilitates recycling of stored water for future uses.

Irrigation water management

There is an urgent need for the judicious use of water from source to fields so the emphasis should be on extensive irrigation rather than on intensive. **Precision irrigation** Computer based sensor is the simple instrument to measure in the field soil moisture for prescribing time and quantity of irrigation. Sprinkler and drip systems of irrigation increased crop yield , save sufficient amount of water and control over-exploitation of ground water. Water use efficiency increased further by adequately fertilizing the crops through drip irrigation (Faroda, *et al*, 2007). **Irrigation applied at the critical stages** of the crops play a paramount role in saving water and hence increased water use efficiency. **Rotational irrigation practice** it is defined as the application of water to field at the regular intervals. It can be successfully applied in the rice field to prevent the drawbacks of the continuous flooding practices in the rice. Under this practice water is not kept standing in the rice fields and irrigation is provided in the intervals but it is taken in to consideration that soil does not become enough dry that it may cause moisture stress. This practice may save water up to 40% as compared to the continuous flooding practices and total requirement of the rice can be reduced to 700 -850mm and enhances water use efficiency from 50to 70% (Bali,2010)

Water saving technology

For water conservation, rice is an important target as rice crop consumes 3000 to 5000 litres of water to produce 1kg of rice and accounts for 50% irrigation water. Aerobic rice is a new concept of growing rice , it refers to high yielding rice grown in non-puddle aerobic soils without making the water to stand. Aerobic rice yields varied from 4.5 to 6.5 tonns per hectare which is about double than that of traditional varieties and about 25 to 35 % lower than that of low land varieties grown under irrigated or flooded conditions(Haradari,2010). aerobic rice has great advantages.

1. It decreased methane gas production which is an important source of global warming.
2. It increase the mycorrhizal association as aerobic conditions provide favourable environment to the mycorrhizal fungi.

3. It also increased rhizobial association, rhizobia in these conditions grow as endophytes in and on the roots and contribute to increase the rice production.
4. It prolonged root activities, as rice plants that are growing in continuously water saturated soils degenerate about 75% of its roots, whereas, in the aerobic conditions there is no degeneration of the roots.
5. It increased nutrient uptake, reduced nitrogen losses and reduced incidence of insect, pests and diseases.

7

Farming System Approaches Towards Livelihood Improvement

Farm economic efficiency of a specific area is an important factor of productivity and growth, where resources are scare and opportunities for developing and adopting better technologies have lately declining. No doubt, conventional farming is risky and farmers are reluctant to invert heavily in crop production. Modest increments in productivity are no longer sufficient to justify the investment of scarce resources as farming sector is small scale and resource poor, but integrated farming system assumes greater importance for sound management of farm resources to enhance the farm productivity, reduce the environmental degradation, improve the quality of life of resource poor farmers and to maintained sustainability.

Farming system refers to the farm wherein two or more enterprises are integrated with the farm resources for achieving the fuller utilization, realize maximum profit and stabilize returns.

It is a resource management strategy to achieve economic and sustained agricultural production to meet diverse requirements of the farm household while preserving the resource base and maintaining high environmental quality. The objectives of farming system in general are converging as to the development of sustainable location specific farm technology to raise and sustain the total farm productivity in terms of food ,feed, fodder and fuel (4F) to meet the felt needs of the farmers within the sphere of their agro-socio-political favourites and constraints. Farming systems aim for increased productivity, profitability, sustainability, balanced food, pollution free environment, recycling of unutilized resources, generation of income round the year, adoption of new technology, solving energy ,fuel and fodder crisis, avoiding deforestation, increased employment generation, literacy rates, input-output efficiency, enhanced opportunity for agriculture oriented industries and standard of living of the farmers.

Some studies on farming system shows that farming system approach is better than conventional farming and main advantages of farming systems include generation of more employment,better spread of labour over a period of time,reducing risk,conservation of resourcews and higher economic returns.In view of the serious limitations on horizontal expansion of land and agriculture,only alternative left is vertical expansion through various farm enterprises required less space and time but giving high productivity and ensuring periodic income especially for small and marginal farmers of hilly area of our regions like Doda, Bhaderwah, Rajouri, Ramban etc.

Generally farming systems practiced by farmers vary according geographical locations,climatological conditions,infrastructure facilities available, socially acceptable,socio-economic conditions of the farmers

Farming system for high hills of Jammu region especially, alpine conditions of himalayas

High hills of Jammu region comprises Doda, Bhaderwah and Kishtwar are located above 1500 meters altitude coming under temperate zone is characterized by mountainous topography. It has diverse agro-climatic condition from subtropical plain to semi-temperate to alpine and sub-alpine conditions. This zone is characterized by relatively mild but dry summer with little monsoon and fairly cold- wet winter due to the 'Western Weather Disturbances', receiving annual rainfall of 900 mm, most of it is received between June to September .The land in this district generally consists of clay loamy soil, clay sandy loam and sandy loam. Clay loamy soils are very suitable for cultivation of apple, pear, peach ,paddy maize where as clay sandy loam is very suitable for the cultivation of walnuts. These soils have the capacity to retain moisture of 18-30% and is suitable for rabi cultivation also.

Doda district comprises Doda, Bhaderwah, Kishtwar and Ramban having area of 11691 km^2 with altitude 750-2500 msl having 32° 53' to 34° 21'N latitude and 75°1' to 76° 47'E Longitude. It has seven tehsils with cultivated area of 84,900 ha. About 20,000 ha area is under horticulture In this track ,more than 75% of land holders belong to marginal farmers with land holding > 1 ha. It is mostly a mono-cropped zone with low production & productivity. Agricultural crops mostly taken are Maize, paddy, wheat, mustard, oats and pulses. However, other agriculture enterprises like pomology(Horticulture crops and vegetables of cucurbitaceae and solanaceae family), apiculture & animal/sheep husbandry are very common and rather supplement the modest agricultural income obtained through arable farming. The area is also suitable for seed production of temperate vegetable crops. Saffron (*Crocus sativus*), possessing innumerable medicinal and aromatic properties is grown in the Kishtwar plateau of Doda district in an area of about 60 ha. Rajmash (Phaseolus vulgaris) possessing flavour and organo-leptic qualities having high market potential is also widely grown in the higher altitudes of this zone .

Constraints of the prevailing farming systems

Socio-economic issues

- Fragmented land holdings,about 74% of the farmers have marginal land holdings (<1 ha)
- Local strains of many pulses such as Rajmash having better quality and taste which sells at premium in the market, many farmers are not ready to adopt improved varieties thus local strains need to be improved

Technological issues

- About 90% area of the district is rainfed therefore,there is need to develop new varieties which have wider adaptability and resistant to drought situations.
- Inadequate availability of the quality seeds and propogated materials of improved varieties of food grains, vegetables and fruit crops and scientific know how of production of these crops.
- Non-availability of the farm inputs (seeds, fertilizers and pesticides) at right time.
- Less remunerative prices in absence of proper marketing and transport facilities.
- Lack of proper feed, fodder and concentrate to milch animals
- Improper market and prices for egg and broilers
- Poor density of dairy Cooperative societies.
- Poor Veterinary/artificial insemination facilities
- Non-availability of day old chicks as no commercial hatchery exist in the areas.
- High altitude (Low possible oxygen pressure) cause severe stress during transportation of chicks.
- Lack of awareness bioscurity/insufficient veterinary aids.

Enhancement in the agricultural productivity through proven technological interventions

For getting higher returns and employment generation the prevailing farming system of the area could calibrize through adoption of modest and latest farm technologies.

- Making available and popularizing the high yielding varieties of cereals and oilseeds, which are cold tolerant and resistant to shattering

- Development of maize and wheat varieties which are resistant to major insect-pest and diseases of the area.
- Introduction of improved production technology, supply of quality seed and planting material (field and horticultural crops in 1st year) and their production by farmers in subsequent years.
- Intensification and diversification of cropping systems, off-season vegetables, organic farming, agro- forestry, orchard management and farm mechanization.
- Up-gradation of cattle, backyard poultry, sheep and goats.
- Introduction of need and opportunity based mushroom cultivation, fishery, apiary and Seri-culture.
- Processing and preservation of the vegetables
- Evaluation of promising cultivars of apple, pears, peach, plum. apricot, almond, walnut and pomegranate
- Emphasis on hygienic practices to improve quality of animal products
- Improvement of animal health through augmentation of nutrition
- Establishing marketing system which will enable remunerative price to the farmers.
- To educate the orchard growers about scientific cultivation of fruit plants for higher yield and income per unit orchard.
- To educate farmers about commercial poultry production for higher prices of eggs/broiler
- Giving more emphasis on the use of INM and IPM practices

Sustainable IFS models for 0.5 ha area for Bhaderwah area for, livelihood improvement

Appropriate integration of sub-systems scientifically in a farming system is essential to overcome the risk factors and in realizing income through out the year from various enterprises.

Components in Proposed Farming system

- **Cropping**

The crop activity in proposed integrated farming system consists of Horticultural crops(40%), Agricultural crops (40%). the horticultural crops include walnut, apple, pear and peach. Seven cropping sequences are rice-oat+berseem, rice-wheat, rice-peas, rice-garlic, maize+cowpea-mustard, maize+potato-oats, maize+rajmash-potato , maize+cucurbits-onion

- **Livestock component**
  1. Milch cows

     Two Jersey crossbred for hilly area with exotic inheritance >62.5%

  2. Backyard Poultry

Twenty numbers of chicks(10 female +10 male)

- Bee Keeping- five hives
- Mushroom cultivation(Dhingri) –Five bags

Respective proposed area under different farming system components

Integrated farming system	Area(ha)
Horticultural crops	0.20
Walnut	0.10
Apple	0.05
Pear	0.025
Peach	0.025
Cropping sequences	0.2 0
rice- berseem	0.025
rice-wheat	0.025
rice-peas	0.025
rice-garlic	0.025
maize+cowpea-mustard	0.025
maize+potato-oats	0.025
maize+rajmash-potato	0.025
maize+cucurbits-onion	0.025
Farming components (dairy+poultry+apiary+mushroom cultivation)	0.10
Total area	0.5

Table 1: Yearwise economics of horticultural component in proposed integrated farming system

Years	Walnut			Apple			Pear			Peach		
	Cost of cultivation (Rs/ha)	Gross returns (Rs/ha)	Net returns (Rs/ha)	Cost of cultivation (Rs/ha)	Gross returns (Rs/ha)	Net returns (Rs/ha)	Cost of cultivation (Rs/ha)	Gross returns (Rs/ha)	Net returns (Rs/ha)	Cost of cultivation (Rs/ha)	Gross returns (Rs/ha)	Net returns (Rs/ha)
1st	15164	-	-	35389	-	-	31489	-	-	52837	-	-
2nd	4847	-	-	9630	-	-	7970	-	-	23192	-	-
3rd	4723	-	-	9610	-	-	8090	-	-	21214	25000	3786
4th	6157	-	-	11840	50000	38160	7930	20000	12070	15352	50000	34648
5th	9241	24800	15559	13725	70000	56575	11090	40000	28910	16357	93750	77393
6th	10004	49600	39596	15246	100000	84754	12611	60000	47389	18491	125000	106509
7th	10404	62000	51596	16898	120000	103102	14483	80000	65517	20227	187500	167273
8th	10776	99200	88424	17715	150000	132285	15300	120000	104700	21786	237500	215714
9th	17116	148800	131684	19146	180000	160854	16731	160000	143269	23330	250000	226670
10th	17435	186000	168565	20028	200000	179972	17613	200000	182387	24924	312500	287576
11th	18605	223200	204595	21853	250000	228147	19108	240000	220892			
12th	21124	248000	226876	23005	300000	276995	18260	280000	261740			
13th	22521	310000	287479	24136	350000	325864	21941	320000	298059			
14th	26765	347200	320435	25289	400000	374711	23094	360000	336906			
15th	27083	372000	344917	26464	450000	423536	24489	400000	375511			
16th	28124	434000	405876									
17th	28443	496000	467557									
18th	31485	558000	526515									
19th	31803	620000	588197									
20th	35883	682000	646117									
Employment generation mandays (1st year) Walnut 93 Apple 220 Pear 210 Peach 325												

Table 2: Economics and employment generation efficiency of agricultural crops for proposed area

Cropping sequence	Cost of cultivation (Rs)	Gross returns (Rs)	Net returns (Rs)	Employment generation of system (mandays/year)
rice-berseem				10
rice	1000	1625	625	
berseem	500	1125	625	
System performance	1500	2750	1250	
rice-wheat				10
rice	1000	1625	625	
wheat	450	700	250	
System performance	1450	2325	875	
rice-field peas				11
rice	1000	1625	625	
field peas	450	725	275	
System performance	1450	2350	900	
rice-garlic				12
rice	1000	1625	625	
garlic	1705	6000	4295	
System performance	2705	7625	4920	
maize+cowpea-mustard				9
maize+cowpea(1:1)	692	1625	933	
mustard	425	575	1501	
System performance	1117	2200	1083	
maize+potato-oats				8
maize+potato(2:1)	770	1830	1060	
oats	450	750	300	
System performance	1220	2580	1360	
maize+rajmash-potato				9
maize+Rajmash(1:1)	800	1900	1100	
potato	1190	2500	1310	
System performance	1990	4400	2410	
maize+cucurbits-onion				10
maize+cucurbits(3:1)	550	1625	1075	
onion	576	3000	2424	
System performance	1126	4625	3499	

Table 3: Yield and employment generation efficiency of different components under integrated farming system

Components	Cost of cultivation (Rs)	Gross returns (Rs)	Net returns (Rs)	Employment generation (mandays/year)
Poultry (20 birds)	1660	14825	13165	30
Dairy (Two cows)	95400	198550	103150	60
Bee keeping (Five hives)				60
1st year	12500	3000	-	
2nd year	2500	6000	3500	
3rd year	2500	9000	6500	
4th year	2500	12000	9500	
5th year and onwards	2500	15000	12500	
Mushroom cultivation (2 months) (Five bags)	300	1000	700	½ /bag

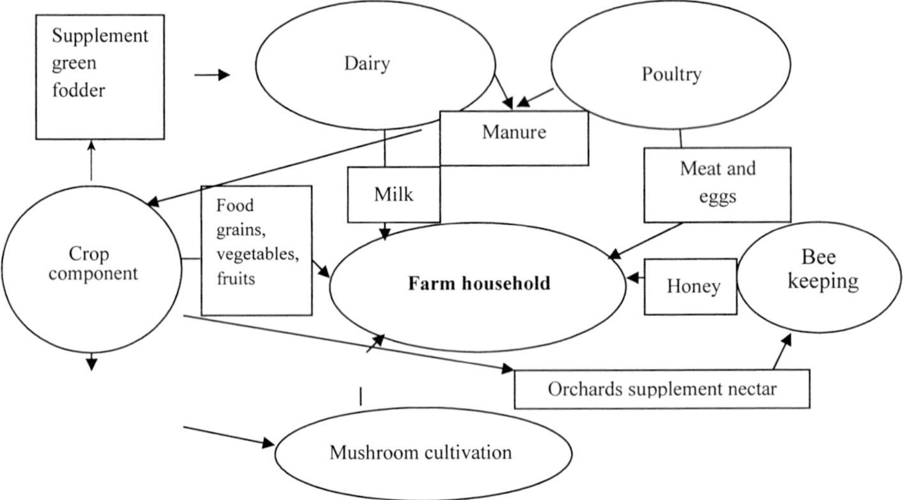

Fig. 1: Resource flow model of integrated farming system-Bhaderwah conditions (0.50ha)

Horticultural crops**(0.20 ha)**	**Agricultural crops (0.20 ha)**	**Mushroom cultivationj (5 bags)**
Gross returns: Rs. 108512 Cost of cultivation: Rs.6146 Net returns: Rs. 102366 Contribution to total income:41.3% Employment generation (mandays):23	Gross returns: Rs 28855 Cost of cultivation: Rs.12558 Net returns: Rs. 16297 Contribution to total income:6.5% Employment generation (mandays):79	Gross returns: Rs. 1000 Cost of cultivation: Rs.300 Net returns: Rs. 700 Contribution to total income:0.28% Employment generation (mandays): ½ manday/bag

Backyard Poultry (20 birds) Gross returns: Rs. 14825 Cost of cultivation: Rs.1660 Net returns: Rs. 13165 Contribution to total income:5.3% Employment generation(mandays)	Bee Keeping (5 hives)*** Gross returns: Rs. 15000 Cost of cultivation: Rs.2500 Net returns: Rs. 12500 Contribution to total income: 5 % Employment generation (mandays): 12	Dairy(2 Cows) Gross returns: Rs.198550 Cost of cultivation: Rs.95400 Net returns: Rs. 103150 Contribution to total income:41.5% Employment generation (mandays): 60

** For Horticultural crops economics of walnut is calculated from 20th year, apple and pear from 15th year and peach 10th year

*** economics of 5th year is taken in to consideration

System Gross income Expenditure Net return
(Rs) (Rs) (Rs)

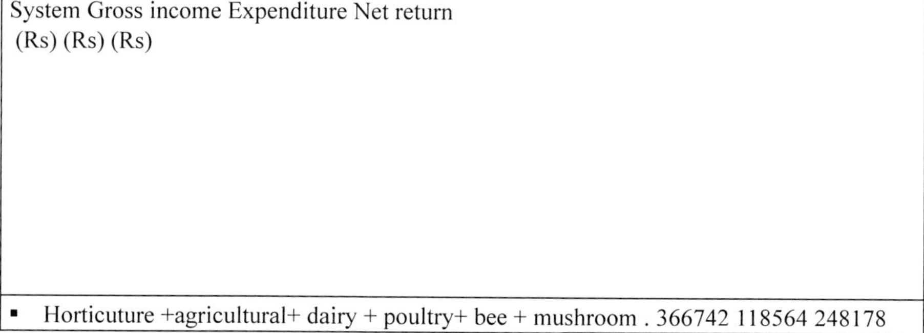

- Horticuture +agricultural+ dairy + poultry+ bee + mushroom . 366742 118564 248178

Fig. 2: Profitability and employment generation in proposed integrated farming system for Bhaderwah region of area 0.50 ha of Jammu and Kashmir

On the whole,the integration of cropping components with different farming system components is able to generate more income and increased the profitability through recycling of wastes of one component into another (Fig.1). This also reduced the dependence on external inputs. The integration of horticulture components with different farming system increased the net returns to the tune of Rs. 248178 (Fig. 2) and also increased employment opportunities and generated enough scope to employ family labourers around the calendar year. However this model is developed for small farmers having holdings of 0.5 ha and the labourer engaged in crop components also take care of dairy, poultry, apiary and mushroom cultivation reducing the cost of production with increased profitability.

References

Ahmad 1984. Soil conservation needs in the hilly areas of Bangladesh.Task force report, resource loss and economic loss in the hill tract areas of Bangladesh. BARI, Report, Joydepur, Bangladesh

Amareswari PU, Sujathamma P. Jeevamrutha as an alternative of chemical fertilizers in rice production. Agricultural Science Digest - A Research Journal. 2014;34(3):240.

Anonymous.2010. Report of the evaluation study on hill area development programme in Assam and West Bengal organized by Planning Commission, Government of India.

Azad,K.C.R Swarup and B.K.Sikka.1988.Horticultural development in hill areas, Mittal publications .Delhi

Bajracharaya,B.B. 1992.Infrastructural development imperatives for sustainable Mountain Agriculture, In N.S. Jodha, M. Banskota and T. Partap(eds),Sustainable Mountain Agriculture ,vol.1 Oxford and IBH Publishing Co., pp.235-252.

Behera, U.K., Jha, K.P. and Mahapatra, I.C. 2004. Integrated management of available resources of the small and marginal farmers for generation of income and employment in Eastern india.Crop Research 27(1):83-89

Bharucha ZP, Mitjans SB, Pretty J. Towards redesign at scale through zero budget natural farming in Andhra Pradesh, India. International Journal of Agricultural Sustainability.2020;18(1):1-20.

Carating R.B., Galanta R.G., Bacatio C.D. (2014) The Soils of the Hills and Mountains. In: The Soils of the Philippines. World Soils Book Series. Springer, Dordrecht

Changkija,S. 2017. Sustainable Hill Agricultural Diversity of Nagaland. International Journal of Tropical Agriculture.35(1):121-133. Tropical

Critchley, W. 1991. Looking After Our Land, Soil and Water Conservation in Dryland Africa, Oxfam, London, 84 p.

Critchley, W.; Siegert, K.; Chapman,C.; Finkel, M. 1991. Water harvesting: A Manual for the Design and Construction of Water Harvesting Schemes for Plant Production, Food and Agriculture Organization of the United Nations, Rome .

Critchley, W.R.S. and Reij, C. (1989). Water harvesting for plant production: Part 2[Case studies and conclusions from Sub-Saharan Africa. (Draft) .To be published in 1991 as the final report of the World Bank's Sub-Saharan Water Harvesting Study. Contains case studies of both traditional techniques and project involvement in WH. The conclusions cover engineering design, production aspects and socio-economic issues.

Driml,Sally and Mick Common.1995. Economical and financial benefits of Tourism in major protected areas". Australian journal of Environmental management 2:19-29.

Duveskog, D. 2001. Water harvesting and soil moisture retention. A study guide for Farmer Field Schools.Ministry of Agriculture and Farmesa, Sida, Nairobi.

Farid, A. T.M. and Hossain,S.M.S. 1988.Diagnosis of farming practice their impact on soil resource loss and economic loss in the hill tract areas of Bangladesh BARI, Report,Joydepur, Bangladesh

Fatima, K. and Hussain, A.2013. Problems and Prospects of Hill Farming. Research Journal of Agricultural Sciences 2012, 3(2): 578-580

Georgis, K. Temesgen,M. and Goda,S.2001.On-Farm evaluation of soil moisture conservation techniques using improved germplasm.7th Eastern and Southern Africa regional Maize Conference held on 11-15th Feb.2001.PP 313-316.

Gill, M.S.,Toor,M.S. and Ramasundaram,P. 2007.Integrated Farming system-a success story. Journal of farming system research and development 13(1),10-16.

Gim, U.H. 1998. Republic of Korea-A country report,In: proceeding of the workshop on perspectives on Sustainable Farming Systems in upland areas.Tokyo, Asian Productivity organization..

Gupta, N.K. and Mahajan,P.2010. Concept of watershed and watershed management. Training manual of fourteen days training of trainers on 'scaling of water productivity in agriculture for livelihood through teaching cum demonstration'organised by water management research centre Directorate of research, SKUAST- Jammu

Gupta, A. and Gupta,N.K.2010. Agronomic practices in a water shed management project .Training manual of fourteen days training of trainers on 'scaling of water productivity in agriculture for livelihood through teaching cum demonstration'organised by water management research centre Directorate of research, SKUAST- Jammu

Gupta, S.K. Agroforestry for Sustainable farming in the hills. Training manual of fourteen days training of trainers on 'scaling of water productivity in agriculture for livelihood through teaching cum demonstration'organised by water management research centre Directorate of research, SKUAST- Jammu

Haradari, Chandrashekhar.2010.Aerobic rice: a water saving technology. Achieving sustainability in agriculture issues challenges and opportunities.Agrobios 9(4):42

Hartley K. and Nicholas,H.1992. Tourism policy: Market failure and Public choice, In Peter Johnson and Barry Thomas (eds),Prespectives on tourism policy,London :Mansell Publishing Ltd., 15-28.

Hudson, N.1987.Soil and water conservation in Semi-arid areas.Bernan press(PA).IWMI Vallerani system.

Islam, M.A.1984. Soil erosion and Conservation in the hilly areas of Bangladesh BARI project report.

Jenkins, C.L. and B.M. Henry.1982. Government involvement in Tourism in developing countries, annals of Tourism research,9(4),499-521.

Jodha, N.S.,banskota,M. and partap,T. 1992. Stratigies for sustainable development of Mountain Agriculture:An overview, In N.S. Jodha banskota,M. and partap,T. (eds),Sustainable Mountain agriculture ,vol. 1 ,oxford and IBH publishing Co., New Delhi,3-40

Jodha. 1990. Mountain perspective framework. Discussion paper,MFS1/1990. www.icimod.org

Joshi PK, Jha AK, Wani SP, Joshi L and Shiyani RL. 2005. Meta-analysis to assess impact of watershed program and people's participation. Research Report 8, Comprehensive Assessment of watershed management in agriculture. International Crops Research Institute for the Semi-Arid Tropics and Asian Development Bank. 21 pp.

Karlberg, L., Rockström, J., Falkenmark, M., 2009. Water resource implications of upgrading rainfed agriculture – focus on green and blue water trade-offs. In Rainfed Agriculture – Unlocking the Potential. S. Wani, J. Rockström, T. Oweis, T. (eds.). Comprehensive Assessment of Water Management in Agriculture. Series Vol. 7. CABI Publication, Wallingford, U.K., pp. 44-57

Khadse A., Rosset P.M., Morales H. and Ferguson B.G. (2018) Taking agroecology to scale: Thezero budget natural farming peasant movement in Karnataka, India. The Journal of Peasant Studies, 45(1):192-219.

Khanday B A, Singh K N, Kanth R H, Chand L, Singh D K and Sofi K A. 2004. Strategies for sustainable crop production in hilly regions of north-western Himalayas. Lecture on Ecological vulnerability and strategies for sustainable agriculture in hilly regions. pp62-65.

Khera, J.S. and Hadda, M.S. 2010.Soil conservation techniques for sustainable crop production in Punjab.Intensive Agriculture 25-26

Liniger,H. and Critchley,w. 2007.Where the land is greener:case studies and analysis of soil and water conservation initative world wide.CTA.

Manandhar,GB, Shrestha,K.B. and Adhikary,S.K.2006. Mechanisation of Strategy to commercialized Agriculture Paper presented at a workshop organized by FNCCI during Nepal Agro Mechanisation and Technology development Epo-2006.March 2-8, 2006. Biratnagar,Nepal.

Meena,V.S. and Sharma,S. 2015. Organic farming: A Case Study Of Uttarakhand Organic Commodity Board .Journal of Industrial pollution control. ISSN (0970-2083)

Miah, M.M.U et al .1993. fertilizer and Soil Management guide for the hilly regions of Bangladesh an unpublished report of BARC,Bangladesh.

Molden D. 2007. Water for Food, Water for Life: A Comprehensive Assessment of Water Management in Agriculture. Molden D (ed). International Water Management Institute IWMI. Earthscan, London, UK; Colombo, Sri Lanka..

Nakagawa,S. 1998. The current situation and future challenges in slope land agriculture in Japan and Asia.In proceedings of the workshop on perspectives on sustainable farming systems in Upland areas, Tokyo, Asian productivity organization.

National Commission on Farmers. 2005. Serving Farmers and Saving Farmers: Crisis to Confidence. Hill Agro-ecosystem, Second Report pp142-228. Ministry of Agriculture. Govt of India.

Nissen,P.,E .2006.Water from roads.Ahanbook for technicians and farmers on harvesting rainwater from roads, Danish International Development Assistence.

Olivier.J.2002.Fog water harvesting along the west coast of South Africa :A feasibility study .water sa-pretoria.,28,349-360.

Pariyar.2001.Baseline study on Agricultural mechanistion needs in Nepal. Rice Wheat consortium for indogangetic plains ,new delhi,india.

Partap, T., 1998 "Sloping Land Agriculture and Resource Management in Semi-arid and Humid Asia: Perspectives and Issues. In Perspectives on Sustainable Farming Systems in Upland areas. Asian Productivity Organization, Tokyo. pp38-84. www.apo.org

Partap, T., 1999. "Sustainable Land Management in Marginal Mountain Areas of the Himalayan Region". Mountain Research and Development, Vol. 19, No.3, pp 251-260.

Partap,T. and Partap,B. 2010.Mountain farmers adaptive stratigies in response to impact of climate change on their livelihood options. In proceedings of the International Symposium on Benefitting from Earthb observation: Bridging the data gap for adaptation to climate change in the HKH region,held during Oct 4-6,2010.Pub.ICIMOD, Kathmandu,Nepal. www.icimod.org

Partap,T. and Vaidya, C.S. 2009.Organic farmers speak on Economics and beyond:A nationwide survey of organic farmers experiences in India.Book published by Westville Publishing House,New Delhi.pp161

Partap,T.1995. High value cash crops in mountain farming:Mountain Development processes and oppurtunities. MFS discussion paper,series No.95/1.ICIMOD,Kathmandu:International centre for integrated mountain development

Quedraogo,E., Mando,A.and Zambre,N.P.2001.Use of compost to improve soil properties and crop productivity under low input agricultural system in west Africa.agriculture ecosystems and Environment,84:259-266

Raju KV, Aziz A, Sundaram MSS, Sekher M, Wani SP and Sreedevi TK. 2008. Guildelines for Planning and Implementation of Wate rshed Development Program in India: A Review. Global Theme on Agroecosystems Report 48. Andhra Pradesh, India: International Crops Research Institute for the Semi-Arid Tropics.

Rao Ram Mohan MS, Chittaranjan S, Selvarajan S and Krishnamurthy K. 1981.Proceedings of the panel discussion on soil and water conservation in red andblack soils, 20 March 1981, UAS, Bangalore, Karnataka: Central and Soil and Water Conservation Research and Training Institute, Research Center, Bellary, Karnataka and University of Agricultural Sciences, Bangalore, India. 127 pp.

Reddy,Y.T. and Reddi,S.G.H.2005.Principles of agronomy by Kalyani Publishers

Rijk,A.G.1989.Agricultural mechanization policy and strategy.Asian productivity organization, Tokyo, Japan

Rockström J, Hatibu N, Oweis T, Wani S, Barron J, Bruggeman A, Qiang Z, Farahani J, and Karlberg L. 2007. Managing Water in Rain-fed Agriculture. In: (Molden D, ed), Water for Food, Water for Life. A Comprehensive Assessment of Water Management in Agriculture. International Water Management Institute. Earthscan, London, UK.

Sarada O. and Kumar G.S. 2018. Perception of the farmers on zero budget natural farming in Prakasam district of Andhra Pradesh. The J. Res. PJTSAU, XLVI(1):34.

Shrestha.1992. a study on indicators of unsustainability of hill farming in Nepal.

Shukla,L.,Singh, Surender and Tyagi, S. 2011.Conservation Agriculture and bio-fertilisers. Indian Farming.5-9.

Singh RV. 2000. (Ed.) Watershed planning and management. Yash Publishing House, Bikaner, Rajasthan, India.

Singh, Kalayan ,Bohra,J.S., Singh, Y. 2007.Development of farming system models for irrigated condition of the North-eastern plain zone of Uttar Pradesh. Journal of farming system research and development 13(1):10-16

Singh, Kalayan ,Bohra,J.S., Singh, Y. and Singh,J.P..2006.Development of farming system models for the north-eastern plain zone of Uttar Pradesh. Indian farming 56 (2):5-11

Smith, J., Yeluripati, J., Smith, P., & Nayak, DR. (2020). Potential yield challenges to scale-up of zero budget natural farming. Nature sustainability, Analysis: 1-6.

Subagyono,K. and Pawitan,H.2008.Water harvesting technique for sustainable water resources management in Catchments area.Proceeding of international workshop on integrated watershed management for sustainable water use in a humid tropical region.Univ. Tsukuba.Japan.

Sugaya,H.1998.Japan.A country report.In :Proceedings of the workshop on perspectives on sustainable farming systems in upland areas.Tokyo,Asian Productivity organization.

Walia, S S and Kaur, Navdeep.2013.An Ecofriendly Approach for Sustainable Agricultural Environment – A Review. Greener Journal of Agronomy, Forestry and Horticulture Integrated Farming System - VL - 1 DOI - 10.15580/GJAFH.2013.1.071813740

Wani SP, Venkateswarlu B, Sahrawat KL, Rao KV and Ramakrishna YS (eds.). 2009. Best-bet Options for Integrated Watershed Management. Proceedings of the Comprehensive Assessment of Watershed Programs in India, 25-27 July 2007, ICRISAT, Patancheru 502 324, Andhra Pradesh, India. Patancheru 502 324, Andhra Pradesh, India: International Crops Research Institute for the Semi-Arid Tropics. 312 pp. ISBN 978-92-9066-526-7: Order code: CPE 167.

Wani, M.H. 2011. Hill Agriculture in India: Problems and Prospects of Mountain Agriculture Ind. Jn. of Agri. Econ. 66(1):64-66.